T0327484

Rheology, Physical and Mechanical Behavior of Materials 1

Series Editor
Noël Challamel

Rheology, Physical and Mechanical Behavior of Materials 1

Physical Mechanisms of Deformation and Dynamic Behavior

Maurice Leroy

WILEY

First published 2023 in Great Britain and the United States by ISTE Ltd and John Wiley & Sons, Inc.

ISTE Ltd
27-37 St George's Road
London SW19 4EU
UK

www.iste.co.uk

John Wiley & Sons, Inc.
111 River Street
Hoboken, NJ 07030
USA

www.wiley.com

Any opinions, findings, and conclusions or recommendations expressed in this material are those of the author(s), contributor(s) or editor(s) and do not necessarily reflect the views of ISTE Group.

Library of Congress Control Number: 2023940623

British Library Cataloguing-in-Publication Data
A CIP record for this book is available from the British Library
ISBN 978-1-78630-765-1

Contents

Preface

In many circumstances, different materials are subjected to high levels of stress, which may result from the processes of forming, shock and impact problems, dynamic stresses of certain structural elements and other causes. Also, in recent years, the interest shown in plastic deformations at high speeds has developed considerably.

The researchers concentrated their efforts in laboratory studies using simple tests, within a speed range from 10^2 to 10^4 s^{-1} in order to fill the gap existing between the tests done on conventional hydraulic machines and those done with explosives. In addition to carrying out inexpensive tests, they were seeking to achieve easy implementation and provide as much information as possible on the behavior of the materials that were tested.

The experimental methods used for these studies vary greatly. Some of them are given in Table P.1, with the range of deformation speeds in which they are implemented.

In this book, three types of actions on materials are used (see Figure P.1): mechanical, electromagnetic and electrohydraulic stresses, $10^2 \leq \dot{\varepsilon} \leq 10^4$ s^{-1}.

$\bar{\dot{\varepsilon}}$ (S^{-4})	Type of device	Experimental difficulties
$< 10^{-4}$	Creep	-
10^{-4} to 1	Mechanical or hydraulic	-
1 to 10^2	Hydraulic or pneumatic	Resonance of the device
10^2 to 5.10^3	Hopkinson bar impact machines	Wave propagation, adiabatic heating
$> 5.10^3$	Impact of projectiles	Wave propagation, high pressures heating
10^3 to 10^6	Expansion of structures caused by explosives	High pressures difficult to measure

Table P.1. *Deformation speed*

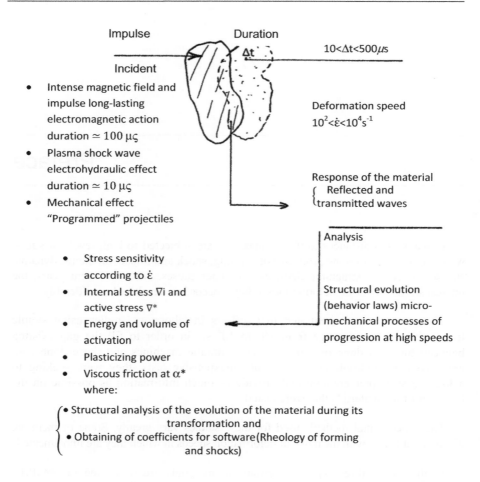

Impulse

Duration

$10 < \Delta t < 500 \mu s$

Incident

- Intense magnetic field and impulse long-lasting electromagnetic action duration $\simeq 100\ \mu\varsigma$
- Plasma shock wave electrohydraulic effect duration $\simeq 10\ \mu\varsigma$
- Mechanical effect "Programmed" projectiles

Deformation speed
$10^2 < \dot{\varepsilon} < 10^4 s^{-1}$

Response of the material
{ Reflected and transmitted waves

Analysis

- Stress sensitivity according to $\dot{\varepsilon}$
- Internal stress ∇i and active stress $\nabla*$
- Energy and volume of activation
- Plasticizing power
- Viscous friction at $\alpha*$
  where:

Structural evolution (behavior laws) micro-mechanical processes of progression at high speeds

{ • Structural analysis of the evolution of the material during its transformation and
  • Obtaining of coefficients for software (Rheology of forming and shocks)

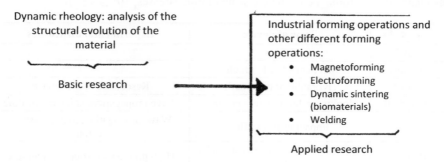

Dynamic rheology: analysis of the structural evolution of the material

Basic research

Industrial forming operations and other different forming operations:
- Magnetoforming
- Electroforming
- Dynamic sintering (biomaterials)
- Welding

Applied research

Figure P.1. *Dynamic stresses – techniques used*

Dynamic Plasticity: Dislocations

1.1. Introduction: how to describe plasticity?

The microscopic description of the plastic strain of metals and alloys mainly relies on the knowledge of the mechanisms for the generation and propagation of dislocations. Though the modeling of macroscopic observations using these microscopic mechanisms has made great progress from a qualitative point of view, much remains to be done from a quantitative point of view.

Under these conditions, the use of a descriptive approach to macroscopic observations remains of great interest. These observations are made on the basis of an examination of mechanical tests in which the strain experienced by the metal as a result of the action of a given system of stresses.

It is necessary here to distinguish the mechanical properties where the time and the rates of strain play only a secondary role (plasticity in the usual sense of the term) from those where time and/or strain rates play an important role (creep, fatigue, dynamic plasticity, etc.).

The mechanical properties reflect the microscopic behavior of the material, and the strain observed at the macroscopic level is the result of local strains on a much finer scale. This microscopic aspect is fundamental for the understanding of different phenomena.

That is why the mechanical tests are often supplemented by a local physical study of the strain mechanism (observations using X-rays, optical microscopes, electronic scanning microscopes, transmission electron microscopes, etc.).

In order to better identify the fundamental mechanisms, it is essential to work on well-defined systems. This is why many studies of the plasticity have been carried out on single crystals.

However, the most commonly used materials are polycrystalline, meaning that they are made up of a more or less isotropic group of monocrystals. It should be noted that for polycrystalline materials, it is more correct to speak of quasi-isotropy than of isotropy.

If the technological material deviates significantly from an isotropic state, then it is said to have a texture. More precisely, a material has a texture if the orientation of its monocrystalline grains, which are generally very numerous, is not totally random but instead presents specific directions which are prevalent. Textures are created at the time the material solidifies, or during an anisotropic strain. They can be transformed by annealing or through phase change. They are of interest economically insofar as they make it possible to improve certain properties of the materials.

Despite the interest industry has in textured materials, we will only consider quasi-isotropic materials in this book.

The application of the experimental laws of plasticity from polycrystalline materials to the calculation of structures in plastic strain, or to the study of processes for forming, can be done by associating two tensors with the stresses and strains shown in the tests with two quantities:

– the generalized stress (or equivalent stress) $\bar{\sigma}$;

– the generalized strain (or equivalent strain) $\bar{\varepsilon}$.

It can then be shown that the curve of the variations of $\bar{\sigma}$ on the basis of those of $\bar{\varepsilon}$ is independent of the type of mechanical test performed. The value of $\bar{\sigma}$ for a given load can thus be deduced from the results of the tensile test, which gives a very particular interest to this type of simple test.

1.1.1. *Equivalence between forming processes and mechanical tests*

A forming method is in itself a mechanical test. Therefore, it is equivalent to any mechanical test conducted under the conditions indicated above.

A number of simple modeling methods allow for a practical implementation of the laws of plasticity.

REMARK.– The forming processes are generally carried out in dynamic plasticity, with strain speeds from 10 to 10^4 s^{-1} (including cases such as forging, wire drawing, stamping, rolling, metal cutting with machining at high speeds, pulsed magnetic fields, explosive, etc.). These cases involve a dynamic rheology, which will be the subject of a dedicated chapter.

1.1.2. *Early stages of strain*

The yield strength is usually defined as the stress above which a strain does not return to zero once the material is no longer subjected to a load.

This problem of the elastic limit is at the origin of the theory of dislocations formulated by Orowan, Polanyi and Taylor, who noted that in order to attain a plastic strain on a material, it is not necessary to deform it over the entirety of its volume. Rather, it suffices to propagate a dislocation line along a slip plane. However, a macroscopic plastic strain may only result from the propagation of a large number of dislocations. Before a deformation, each crystal contains within itself an initial network of dislocations, known as the Frank network, though there is only a small number of these dislocations.

Thus, it will only be possible to observe a macroscopic plastic strain if the dislocations of the Frank network multiply. This can occur in ways such as through the Frank–Read mechanism.

1.1.3. *Multiplication of dislocations*

The initial density of dislocations present in a crystal does not give rise to a plastic deformation of very large amplitude. For example, let us consider a grain of 100 μm on each side, containing 10^{-8} cm^{-2} dislocations of Burgers vector 2.5×10^{-8} cm. Provided that they stretch across the entire section of the grain, we would have, at most, a plastic strain given by the formula:

$$\gamma = \varrho_D b l_D = 10^8 \times 2.5 \times 10^{-8} \times 10^{-2} = 2.5\%$$

We now need to imagine a mechanism capable of multiplying dislocations as can be observed experimentally. Frank and Read have proposed a reel-like mechanism. Let us suppose that there is a dislocation anchored for various reasons at two points, A and B, distant from L (Figure 1.1).

Under the effect of the split exerted on the slip plane, the dislocation occurs. It adopts a curvature such that an arched element $d\ell$ is in equilibrium under the effect of the force f = τb, and the line tension t is assumed to be constant regardless of the nature of the dislocation.

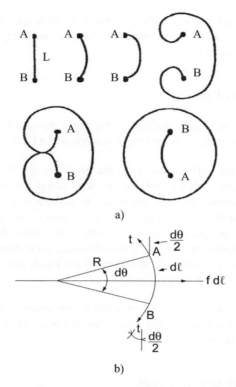

a)

b)

Figure 1.1. *Source mechanisms. (a) Successive positions of a dislocation anchored at A and B and subjected to an increasing force τb (source: Frank–Read). (b) Equilibrium of a dislocation arc*

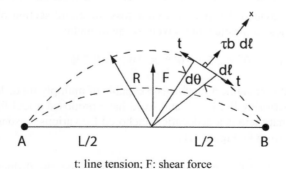

t: line tension; F: shear force

Figure 1.2. *Dislocation arc*

At the equilibrium of the arc of radius R, and knowing that $t = \frac{1}{2}\mu b^2$, we have $|\vec{b}|$ as the Burgers vector and μ as the shear modulus.

$$\sum \text{Forces}/x = \tau b \, d\ell - 2t \sin \frac{d\theta}{2} = 0$$

And, with a small value of $d\theta$:

$$\tau b \, d\ell \simeq 2t \frac{d\theta}{2}$$

It gives:

$$\tau = \frac{t}{b} \frac{d\theta}{d\ell}, \text{ or } Rd\theta = d\ell$$

where:

$$\tau = \frac{t}{b} \cdot \frac{1}{R} \text{ with } t = \frac{1}{2}\mu b^2$$

$$\tau = \frac{1}{2} \frac{\mu b^2}{b} \cdot \frac{1}{R} = \frac{\mu b}{2R}$$

For:

$$R = L/2: \tau_{FR} = \frac{\mu b}{L}$$

τ_{FR} is known as the activation stress of the Frank–Read emitting source.

When τ increases, R decreases until it reaches L/2. The dislocation arc then becomes unstable and rotates around the points A and B, which are distant from L; a recombination of the positions with respect to opposite signs takes place. This results in the formation of a loop which propagates in the slip plane and a new arc AB, which can repeat the same operation.

The multiplication of the number of dislocations can be shown to be done without any difficulty, and therefore it does not control the appearance of the plasticity threshold. This is determined by the difficulty experienced by the dislocations i propagating the crystal. In a pure single crystal, the resistance to the propagation of an isolated dislocation results from its interaction with the crystal lattice. This is the Peierls–Nabarro stress. This stress is very low in the case of cubic face-centered (CFC) monocrystals. It is decisively higher for cubic centered (CC) monocrystals. When these materials are pure, the plasticity threshold is determined by the interactions of the dislocations.

Each dislocation is effectively a distortion of the crystal lattice, which tends to oppose the movement of the other dislocations, in particular those which are perpendicular to it. The presence of impurities and grain boundaries strengthens the resistance to propagation that is exerted on mobile dislocations.

Figure 1.3. *Growth fronts on the surface of electrolytically polished aluminum samples: these fronts are surface steps that result from the emergence of two screw dislocations (Marchin and Wyon 1962; Acta. Met., 10, 915)*

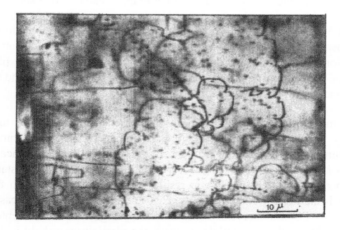

Figure 1.4. *Kimberlite nodule, an example of anchoring and the formation of arcs of dislocations (Gueguen 1979)*

1.1.4. *Fine-level observations (load–unload cycles) of the beginning of plasticization*

An observation at a fine level shows that the displacement of the dislocations begins from the elastic domain. The observation technique used consists of carrying out successive load–unload cycles and in measuring the strains. For CFC crystals, at this level of observation, four domains can be defined:

– the elastic domain;

– the anelastic domain;

– the microplastic domain;

– the macroplastic domain.

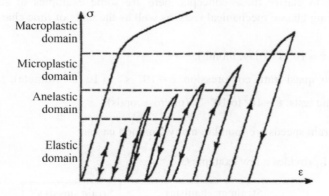

Figure 1.5. *Load–unload cycles*

In the elastic domain, the cycle has a width of zero. The total strain is the sum of the pure elastic strain $\varepsilon_{él}$ and the strain due to the movement of dislocations ε_{dis}:

$$\varepsilon = \varepsilon_{él} + \varepsilon_{dis}$$

1.2. Strain speed: $\dot{\varepsilon}$ ($\dot{\gamma}$ for shearing)

1.2.1. *A few definitions and orders of magnitude*

First, it is desirable to recall a few definitions and orders of magnitude.

In the case of a mechanical test where the force is applied along a single axis, the strain speed is defined by:

$$\dot{\varepsilon} = \frac{d\varepsilon}{dt} = \frac{1}{l_0}\frac{dl}{dt} = \frac{V}{l_0}$$

where l_0 is the initial length of the test piece and V is the difference in the speeds of the ends of the test piece $\dot{\varepsilon}$ (expressed in s^{-1}).

In reality, this definition is sometimes difficult to remember. For example, in the case of a structure subjected to an impact, each point of this structure will be stressed at different speeds, and therefore at different strain speeds. Therefore, the strain speed is no longer homogeneous.

In order to clarify these concepts, here are some examples of strain speeds attained during classic mechanical tests, as well as the units of time characteristic of these tests:

– creep: $\dot{\varepsilon} \approx 10^{-7}s^{-1}$ (day, month);

– traction, quasi-static compression: $\dot{\varepsilon} = 10^{-5}\ s^{-1}$ to $10^{-2}s^{-1}$ (minute);

– dynamic tests: $\dot{\varepsilon} = 10^2$ to $10^4\ s^{-1}$ (microseconds).

These strain speeds are found in many common cases.

Table 1.1 provides a few examples.

Strain mechanisms	Strain speeds s^{-1}
Rolling	1 to 10^3
Spinning	1 to 10^3
Forging	10^{-1} to 5×10^2
Deep pressing	1 to 10^2
Punching	10^{-4} to 10^4
Structural impact (qq m/s)	–
Point of impact	10^2 to 10^3
Other	10^{-1} to 10
Explosion of envelopes (explosives)	10^3 to 10^6
Impact of projectiles	$\rightarrow 10^9$
Adiabatic shear	$>5 \times 10^3$
Crack growth	$>10^4$

Table 1.1. *Strain mechanisms and their speeds*

1.2.2. *Influence of the strain speed*

From many points of view, this influence is equivalent to that of temperature: an increase in the strain speed has the same effects as a decrease in temperature. The sensitivity of the stress to the strain speed at a constant temperature is commonly defined by the following coefficients:

$-\lambda = \left(\frac{\partial \tau}{\partial \ln \dot{\gamma}}\right)_T$ relating to a law of the form $\tau = \tau_y + \ln \dot{\gamma}$;

$-m = \left(\frac{\partial \ln \tau}{\partial \ln \dot{\gamma}}\right)_T$ known as the logarithmic sensitivity to speed, relating to a law of the following type;

$\tau = K\dot{\gamma}^m$.

CFC materials have a yield strength that is not very sensitive to the strain speed (the coefficient m is very low, at 0.014 for copper, for example, Regazzoni 1983). The work hardening increases with the strain speed.

On the other hand, CC materials have an elastic limit that is more sensitive to the strain rate and with consolidation that is influenced little.

HC materials, such as zinc and magnesium, exhibit behaviors similar to CFCs, while the behavior of titanium and zirconium is comparable to CCs.

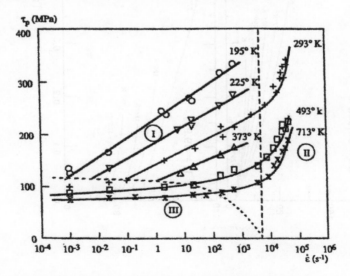

Figure 1.6. *Effect of the strain rate and temperature on the shear yield strength of a mild steel (Campbell et al. 1970b)*

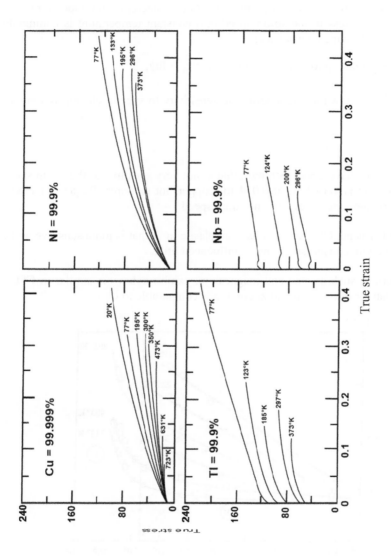

Figure 1.7. *Effect of temperature on the stress–strain curves of various metals (Zackay 1965)*

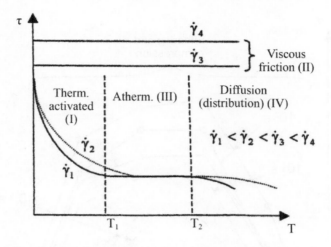

Figure 1.8. *Schematic evolution of the stress with temperatures for different speeds of deformation (Ansart et al. 1991)*

EXAMPLE 1.1.–

Comparison between quasi-static ($\dot{\varepsilon} = 10^{-3}$ s^{-1}) and dynamic (10^2 and 10^3 s^{-1}) strain speeds with the changes in shear strength at $\varepsilon = 10$ and 20% with the different precipitation states of alloy AU$_4$.

1.2.3. *The influence of the "history" of strain*

We have just seen the influences of the temperature and the strain rate on the yield strength and the work hardening, while considering these parameters as constant during the different loads. However, during the plastic strain, the microstructure of the material evolves in ways that differ according to these parameters.

Indeed, if a metal is subjected to a strain rate $\dot{\gamma}_1$, and that the stress is followed by a jump in speed of up to $\dot{\gamma}_2$, the material will respond in a way that is different from that obtained at constant speed $\dot{\gamma}_2$.

The same is true during a sudden change in temperature at a constant deformation speed. This is called the strain history.

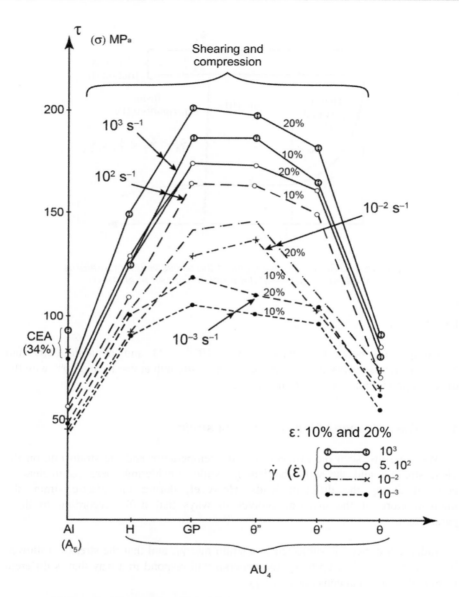

Figure 1.9. *Evolution of the hardening with the precipitation states of AU₄, influence of strain velocities (Leroy et al. 1979)*

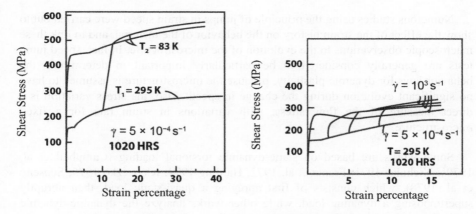

Figure 1.10. *Stress–strain curves for hot-rolled 1020 steel. Effect of the jump in temperature (Hartley 1983). Effect of a jump in speed on strain (Klepaczko 1982)*

The quantitative representation of this effect of the strain history is $\Delta\tau_h$. Several types of sensitivities to the deformation rate are also defined (Klepaczko et al. 1986):

– the instantaneous sensitivity of the stress: $\beta_s = \left(\dfrac{\Delta\tau_s}{\ln\dot{\gamma}_1/\dot{\gamma}_2}\right)_T$;

– the sensitivity of "work hardening": $\beta_h = \left(\dfrac{\Delta\tau_h}{\ln\dot{\gamma}_1/\dot{\gamma}_2}\right)_T$;

– the apparent sensitivity of the stress: $\beta = \left(\dfrac{\Delta\tau}{\ln\dot{\gamma}_1/\dot{\gamma}_2}\right)_T$.

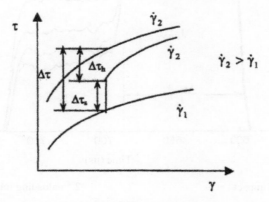

Figure 1.11. *Schematic stress–strain curves which show the effect of the strain history*

Numerous studies using the principle of jumps in strain speed were carried out to show the effect of the strain history on the behavior of the material, and to link these macroscopic observations to the evolution of the microstructure. In fact, speed jump tests are generally considered to be particularly important in determining the behavior laws for dynamic plasticity, because the microstructure is assumed to have no significant evolution during the change in speed. Also, the stress variation is a direct measure of the flow stress, with variations in strain rate for a fixed microstructure.

Some works are based on static-dynamic torsional loading (Campbell et al. 1970a; Nicholas 1971; Frantz et al. 1972; Harding 1974; Klepaczko 1975; Senseny et al. 1978), which consists of first applying a quasi-static load, then abruptly superimposing a dynamic load, while other works analyze the dynamic-dynamic unloading under torsion (Lipkin et al. 1978; Gilat et al. 1988).

1.2.4. *Dynamic–dynamic speed jump*

When the wave train (first impact) reaches the sample, it is impacted by an initial impulse J1 (Figure 1.12).

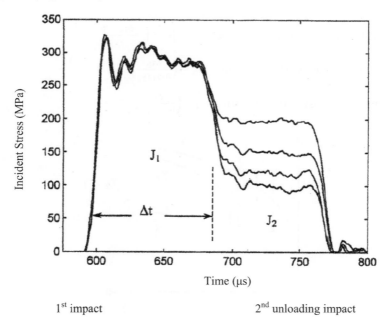

Figure 1.12. *Incident impulses J1 and J2 (Leroy)*

During this impulse, the material is deformed at a speed $\dot{\varepsilon}_1$ up to a plastic deformation ε related to the duration of the impulse. After a time Δt, a second weaker impulse J2 unloads the sample (2^e shock).

The material responds with a decrease in the strain rate from $\dot{\varepsilon}_1$ to $\dot{\varepsilon}_2$ and a decrease in the stress known as a dynamic–dynamic stress jump $\Delta\sigma$ (Figure 1.13).

This response is linked to the microstructural state of the material at the moment the load is released, and therefore to the history of the strain.

EXAMPLE 1.2.– The case of iron $\alpha(cc)$ in an annealed state.

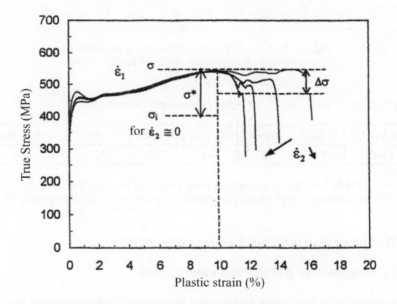

Figure 1.13. *Successive unloading on the same microstructural state (Canto 1998). ε = 10 %, $\dot{\varepsilon}_1$ = 1,400 s^{-1}, iron α in an annealed state. Obtaining the internal stress for $\dot{\varepsilon}_2 \cong 0$*

By taking the impacts J2 for which we have obtained $\dot{\varepsilon}_2 \cong 0$ experimentally, we may consider that we are close to the plasticization limit (Figure 1.13).

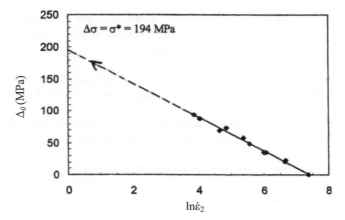

Figure 1.14. *Example of extrapolation: the case of annealed iron for the state 1600-10 (Canto 1998)*

$\sigma_1 = \sigma - \sigma^* = 600 - 194 = 406$ MPa (isothermal values)

State $\dot{\varepsilon}$ (s^{-1})	R-1220	R-1400	R-1600	R-2500	E-1600	E-1890	E-2500
σ^* (MPa)	187	191	194	326	152	161	212

Table 1.2. *Iron: a summary of the values of the effective stress for the different series of tests (annealed state R, work-hardened state E)*

1.3. The microstructural mechanisms of plasticity

1.3.1. *Description of a dislocation and its line*

Formation of a dislocation and its line in a crystal lattice: a crystal half-plane (Figure 1.15(a)) is introduced into the lattice on either side of the cut (Figure 1.15(b)). The new deformed network is illustrated in Figure 1.15(c); the structure of the network is almost perfect, except in the area near the line L. This line L within the network is a dislocation line.

In order to describe the nature of the line L, we use the Burgers vector $\vec{b}$ (a concept introduced as part of the theory of dislocations by J.M. Burgers).

In a plane perpendicular to the line L, a closed circuit (a) is made following a direction of travel, the starting atom 1 and the arrival atom 19 are included within the circuit, which does not include the dislocation. In (b), there is a defect in the closure of the circuit between 19 and 1 in the presence of a dislocation.

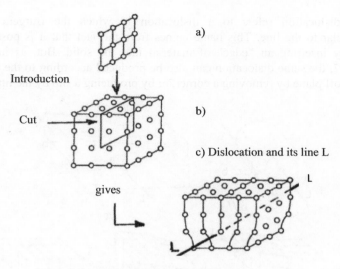

Introduction

Cut

a)

b)

c) Dislocation and its line L

gives

Figure 1.15. *Dislocation*

We define the Burgers vector $\vec{b}$ as being this defect in the closure.

The vector is oriented from the end point of the circuit to the starting point.

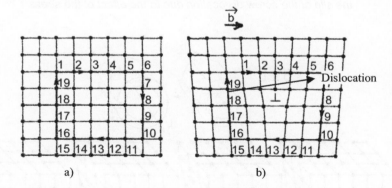

a)

b)

Figure 1.16. *Circuits (a) without and (b) with dislocation*

1.3.2. *Dislocation types*

Screw dislocation refers to a dislocation in which the Burgers vector is parallel to the line. This name comes from the helical transformation of a circuit surrounding the dislocation (Figure 1.17).

Edge dislocation refers to a dislocation in which the Burgers vector is perpendicular to the line. This name comes from the fact that it is possible to be attained by inserting an "edge" of material into the solid. But, as indicated in Figure 1.17, the same dislocation can also be produced according to the orientation of the cut-off plane by removing a corner, or by producing a slip by the lips.

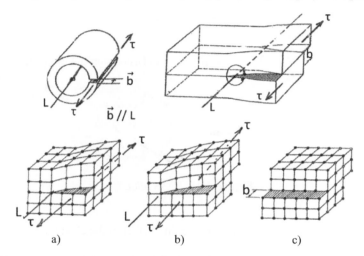

a)   b)   c)

Figure 1.17. *A Burgers vector parallel to the line L, and in (a), (b) and (c), the slip of the screw dislocation due to the effect of the stress τ*

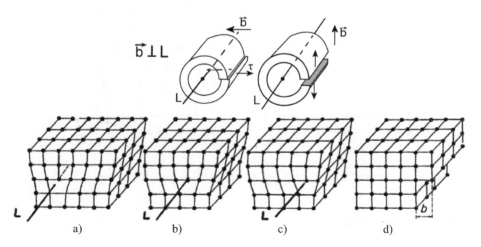

a)   b)   c)   d)

Figure 1.18. *Burgers vectors perpendicular to the line L, and in (b), (c) and (d), the slip of the edge dislocation due to the effect of the shear stress τ*

NOTE. The Burgers vectors of a parallel edge dislocation and a screw dislocation are perpendicular.

Elements	Crystalline system	r_s (Å)	b (Å)	Elements	Crystalline system	r_s (Å)	b (Å)
Li.........	CC	1.72	3.04	Nb.........	CC	1.63	2.86
Be.........	HC	1.24	2.28	Mo.........	CC	1.55	2.72
B.........	R ?	1.22	-	Tc.........	-	-	-
C {d....	CD	1.29	2.52	Ru.........	HC	1.48	2.70
{g....	H	1.29	2.46	Rh.........	CFC	1.49	2.69
Na.........	CC	2.10	3.71	Pd.........	CFC	1.54	2.75
Mg.........	HC	1.76	3.21	Ag.........	CFC	1.59	2.89
Al.........	CFC	1.58	2.86	Cd.........	HC	1.73	2.98
Si.........	CD	1.68	3.84	In.........	TFC	1.84	3.24
P.........	O	1.89	3.32	Sn.........	TC	1.86	3.18
S.........	O	1.83	8.33	Sb.........	R	1.93	4.31
K.........	CC	2.61	4.63	Te.........	H	2.01	4.46
Ca.........	CFC	2.16	3.94	I.........	O	2.17	4.35
Sc.........	CFC	1.93	3.21	Cs.........	CC	3.04	5.25
Ti.........	HC	1.62	2.95	Ba.........	CC	2.47	4.35
V.........	CC	1.52	2.63	La.........	HC	2.07	3.76
Cr.........	CC	1.42	2.50	Hf.........	HC	1.76	3.21
Mn.........	C	1.43	-	Ta.........	CC	1.64	2.86
Fe.........	CC	1.40	2.48	W.........	CC	1.56	2.74
Co.........	HC	1.38	2.51	Re.........	HC	1.51	2.76
Ni.........	CFC	1.38	2.49	Os.........	HC	1.49	2.73
Cu.........	CFC	1.41	2.56	Ir.........	CFC	1.50	2.71
Zn.........	HC	1.53	2.66	Pt.........	CFC	1.53	2.77

Ga.........	O	1.67	3.48	Au.........	CFC	1.59	2.88
Ge.........	CD	1.74	4.01	Hg.........	R	1.76	3.01
As.........	R	1.73	3.76	Tl.........	HC	1.89	3.46
Se.........	H	1.87	4.36	Pb.........	CFC	1.93	3.50
Rb.........	CC	2.81	4.88	Bi.........	R	2.03	4.53
Sr.........	CFC	2.36	4.31	Po.........	C	-	3.35
Y.........	HC	1.97	3.67	At.........	-	-	-
Zr.........	HC	1.71	3.23	Fr.........	-	-	-
				Ra.........	-	2.62	-
				Ac.........	-	-	-
				Th.........	CFC	1.99	3.60
				Pa.........	-	-	-
$1Å = 10^{-10}m$				U.........	O	1.71	2.86

Abbreviations.– C, cubic; CD, cubic, diamond-type; CC, centered cubic; CFC, centered faces and cubic; H, hexagonal; HC, hexagonal compact; O, orthorhombic; T, tetragonal; TC, tetragonal centered; TFC, tetragonal with faces centered; R, rhombohedral; M, monoclinic.

Table 1.3. *Burgers vectors r, the radius of the atomic sphere, and b, the shortest Burgers vector, are given in angströms*

1.3.2.1. Density of dislocations

State	ρ_D (cm^{-2})
Single crystals, carefully solidified	10^2 to 10^3
Annealed single crystals	10^5 to 10^6
Annealed polycrystals	10^6 to 10^7
Extremely work-hardened polycrystals	10^9 to 10^{10}

Table 1.4. *Approximate densities ϱ_D of dislocations*

EXAMPLE 1.3.– Work hardening and dislocation by electromagnetic shocks, copper crystals.

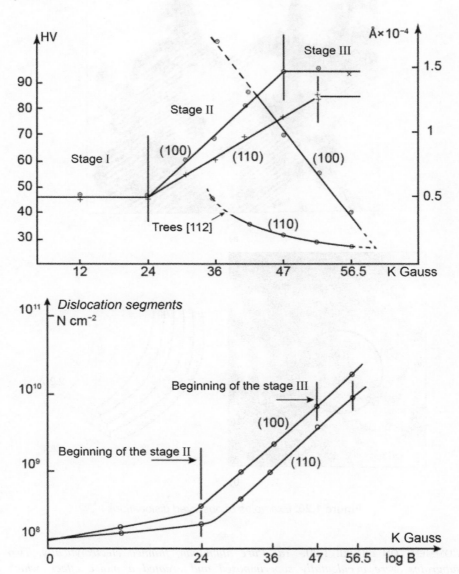

Figure 1.19. *Work hardening, size of the dislocation cells, and density N of dislocation segments in the case of copper crystals [100] and [110] strained at 5 K by electromagnetic shocks (Leroy 1970a)*

1.3.2.2. *Electronic micrographs*

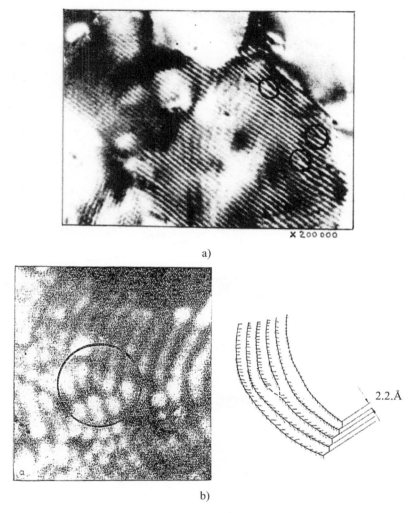

Figure 1.20. *Examples of observed dislocations*

COMMENT ON FIGURE 1.20.– *(a) Pure aluminum, thinned electrolytically. Two subgrains were accidentally superimposed and created a moiré effect, which reproduces the pileup of the reticular planes and their defects (dislocation – surrounded by circles) (Saulnier and Trillat 1962). (b) Micrograph of Mo containing a visible dislocation in the circle, crystallographic planes at a distance of 2.2.Å (Brandon and Wald 1961; Phil. Mag., 6, 1035).*

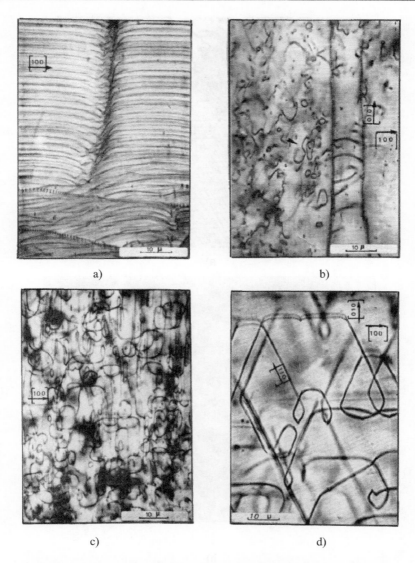

a)   b)

c)   d)

Figure 1.21. *Rock: nodule dynamics. Examples of dislocations observed in rock (nodule of peridotite from the Earth's mantle, basalt with prophyroclastic and protogranular structures) (photos: Guéguen 1976)*

COMMENT ON FIGURE 1.21.– *(a) Screw dislocation [100] moving away from a flexible wall. (b) Loops in the plane [001] and flexible walls [100]. (c) Slips and dipole loops. (d) Multiple slips on [001].*

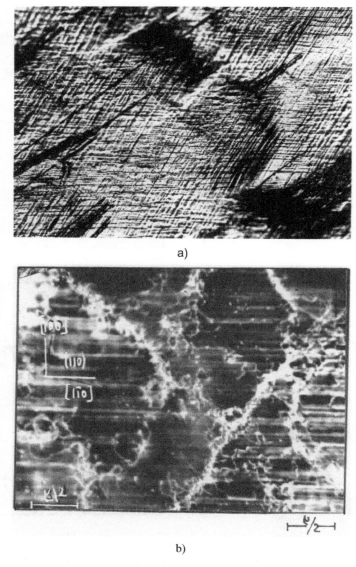

a)

b)

Figure 1.22. *Electromagnetic shocks, 15 μs, $\dot{\varepsilon} \simeq 10^4 s^{-1}$*

COMMENT ON FIGURE 1.22.– *(a) Micrographic aspect of the plane [111]. Lead crystal strained at 5 K by a magnetic field of 4.7 Wb m^{-2}. (b) Electronic micrography (negative) of a crystal of Al [110], stressed by an electro-magnetic* *pulse of 4.7 teslas in ≃ 15 µs. flow [100], with the presence of wall dislocation cells [112] and dislocations on the move, according to [1̄10] during the observation (Leroy 1972).*

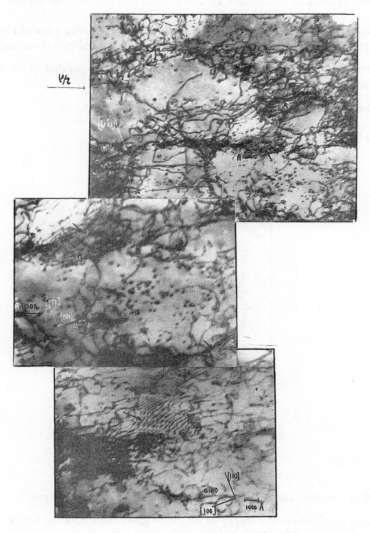

Figure 1.23. *Electromagnetic shocks, ≃ 8 µs, cell walls. Grouping into walls following [100] dislocations. Large number of loops; case of Al strained at 4.7 teslas T = 293 K (Leroy 1970b)*

1.3.2.3. *Stresses, displacement field and strains of a dislocation*

1.3.2.3.1. Stationary screw

The introduction of dislocations creates stresses and strains in the crystal.

Consider a tube of matter at rest. We now cut this tube along a radial plane 0xz. We will strain it elastically, so as to simulate the strains caused by a dislocation:

– If the strain is produced by a translational movement parallel to 0z, a screw dislocation is simulated.

– If the strain is parallel to 0x, an edge dislocation is simulated.

Other inclined strains are possible.

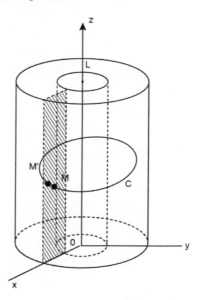

Figure 1.24. *Cut*

The displacements caused $\vec{u}$ are defined by the following conditions:

– Elastic equilibrium: the theory of elasticity gives the following vector relation for displacements $\vec{u}$ of the elastic balance:

$$(1 - 2v) \, \Delta\vec{u} + \overrightarrow{\text{grad div } \vec{u}} = 0$$

v is Poisson's ratio.

– Ratio of the Burgers vector; let M and M' be two points along the cut merged when the body is at rest, and C a Burgers circuit surrounding the dislocation:

$$u(M') - u\,(M) = \int_C \frac{\partial u}{\partial s}\,ds = b$$

If the tube is not subjected to any external stress, the internal stresses must have zero components on the surface of the tube.

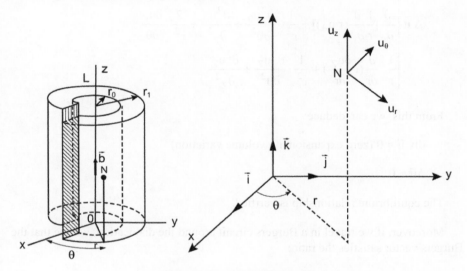

Figure 1.25. *Displacement of the facets*

$\vec{b}$ is parallel to $0z$: due to the symmetry of the system, we use cylindrical coordinates r, θ, z. For a displacement parallel to $0z$, the coordinates of the displacement $\vec{u}$ (u, v, w) of a point N $(r, \theta\ z)$ are:

$$u_\theta = 0,\ u_r = 0,\ u_z = \frac{b\theta}{2\pi}$$

$$(\text{or: } u = v = 0 \text{ and } w = \frac{b\theta}{2\pi})$$

In cylindrical coordinates, we have:

$$\mathrm{div}\ \vec{u} = \frac{1}{r} \cdot \frac{\partial(r u_r)}{\partial_r} + \frac{1}{r} \cdot \frac{\partial u_\theta}{\partial\theta} + \frac{\partial u_z}{\partial z}$$

$$\Delta\vec{u}\begin{cases} \dfrac{\partial}{\partial_r}[\dfrac{1}{r} \cdot \dfrac{\partial}{\partial_r}(r\,u_r)] + \dfrac{1}{r^2} \cdot \dfrac{\partial^2 u_r}{\partial\theta^2} + \dfrac{\partial^2 u_r}{\partial z^2} - \dfrac{1}{r^2} \cdot \dfrac{\partial u_\theta}{\partial\theta} \\[2mm] \dfrac{\partial}{\partial_r}[\dfrac{1}{r} \dfrac{\partial}{\partial_r}(r\,u_\theta)] + \dfrac{1}{r^2} \cdot \dfrac{\partial^2 u_\theta}{\partial\theta^2} + \dfrac{\partial^2 u_\theta}{\partial z^2} + \dfrac{2}{r^2} \cdot \dfrac{\partial u_r}{\partial\theta} \\[2mm] \dfrac{1}{r} \cdot \dfrac{\partial}{\partial r}(r\dfrac{\partial u_z}{\partial r}) + \dfrac{1}{r^2} \cdot \dfrac{\partial^2 u_z}{\partial\theta^2} + \dfrac{\partial^2 u_z}{\partial z^2} \end{cases}$$

From this, we can deduce:

div $\vec{u}$ = 0 (zero expansion, no volume variation)

$\Delta\vec{u} = 0$

The equilibrium relationship is verified.

Moreover, if we travel in a Burgers circuit around the dislocation, we see that the Burgers vector satisfies the ratio:

$$b = \int_C \frac{\partial u}{\partial s}\, dS = u$$

The planes perpendicular to the axis of the dislocation are transformed into helicoids.

1.3.2.3.2. Deformations

Consider the case of a pure dislocation screw. The coordinate system is arranged so that the z axis coincides with the dislocation. We take the vector tangent at the location in the direction of the positive z values. The Burgers vector $\vec{b}$ of the dislocation can be written $\vec{b} = b\vec{k}$, where $\vec{k}$ is the unit vector in the z direction. It can be seen that if the screw is to the left, b is positive, while a negative value of b indicates a screw to the right. It can be reasonably assumed that the displacement in the z direction is:

$$w = -\frac{b}{2\pi}\,\mathrm{arc\ tg}\,\frac{y}{x} = -\frac{b\theta}{2\pi}$$

If we create a circuit around the dislocation line, the arc function tg varies by 2π, and therefore w varies by - b, where u = v = 0.

The elastic strains around the screw dislocation are, in x, y, z coordinates:

$$\varepsilon_{xz} = \frac{\partial u}{\partial z} + \frac{\partial w}{\partial x} = \frac{\partial w}{\partial x} = \frac{b}{2\pi} \frac{y}{(x^2+y^2)^2}$$

$$\varepsilon_{yz} = \frac{\partial u}{\partial z} + \frac{\partial w}{\partial x} = \frac{\partial w}{\partial y} = -\frac{b}{2\pi} \frac{x}{(x^2+y^2)^2} \qquad [1.1]$$

$$\varepsilon_{xx} = \varepsilon_{yy} = \varepsilon_{zz} = \varepsilon_{xy} = 0$$

In cylindrical coordinates r, θ, z, given that:

$$r^2 = x^2 + y^2, \text{ tg } \theta = y/x, z = z$$

and that the elastic displacements parallel to the directions r, θ, z are u_r, u_θ and w, respectively; the elastic strains are given as:

$$\varepsilon_{rr} = \frac{\partial u_r}{\partial r}, \ \varepsilon_{\theta\theta} = \frac{1}{r} \frac{\partial u_\theta}{\partial \theta}$$

$$\varepsilon_{zz} = \frac{\partial w}{\partial z}, \ \varepsilon_{\theta z} = \frac{1}{r} \frac{\partial w}{\partial \theta} + \frac{\partial u_\theta}{\partial z}$$

$$\sigma_{rz} = \frac{\partial u_r}{\partial z}, + \frac{\partial w}{\partial r}, \ \varepsilon_{r\theta} = \frac{\partial u_\theta}{\partial r} - \frac{u_\theta}{r} + \frac{1}{r} \frac{\partial u_r}{\partial \theta}$$

For the screw dislocation, we obtain:

$$w = -\frac{b\theta}{2\pi}, u_r = u_\theta = 0$$

with:

$$\varepsilon_{\theta z} = -\frac{b}{2\pi r}, \varepsilon_{rr} = \varepsilon_{\theta\theta} = \varepsilon_{zz} = \varepsilon_{r\theta} = \varepsilon_{rz} = 0 \qquad [1.2]$$

For $r \to 0$, $\varepsilon_{\theta z} \to \infty$, at the core of the strain, the strains cannot be infinite, and the relationships concerning ε_{xz}, ε_{yz} and $\varepsilon_{\theta z}$ are valid only for distances $r > 5b$ (our analysis is no longer valid in regions close to the dislocation line).

1.3.2.3.3. Stresses

For an isotropic material, the elastic constants are the coefficient μ (shear modulus), the Young's modulus E and Poisson's ratio ν.

The number of constants is reduced to the two Lamé coefficients μ and λ.

Recall that the Young's modulus, the ratio of the longitudinal stress to the longitudinal deformation of a specimen stretched in a single direction, is λ. $(3\lambda + 2\mu)/(\lambda+\mu)$, and that Poisson's ratio, the ratio of the lateral contraction to the longitudinal extension of a specimen stretched in a single direction, is $\lambda/2\,(\lambda+\mu)$.

In the x, y, z coordinate format, we can write the stresses in the form of:

$$\sigma_{xx} = (\lambda + 2\mu)\,\varepsilon_{xx} + \lambda\varepsilon_{yy} + \lambda\varepsilon_{zz}$$

$$\sigma_{yy} = \lambda\varepsilon_{xx} + (\lambda + 2\mu)\,\varepsilon_{yy} + \lambda\varepsilon_{zz}$$

$$\sigma_{zz} = \lambda\varepsilon_{xx} + \lambda\varepsilon_{yy} + (\lambda + 2\mu)\,\varepsilon_{zz}$$

$$\sigma_{yz} = \mu\varepsilon_{yz}$$

$$\sigma_{zx} = \mu\varepsilon_{zx}$$

$$\sigma_{xy} = \mu\varepsilon_{xy}$$

Taking into account the relationships obtained in equation [1.1] for the strains, we therefore obtain stresses equal to:

$$\sigma_{xz} = \frac{\mu b}{2\pi}\frac{y}{x^2 + y^2}$$

$$\sigma_{yz} = -\frac{\mu b}{2\pi}\frac{x}{x^2 + y^2}$$

$$\sigma_{xx} = \sigma_{yy} = \sigma_{zz} = \sigma_{xy} = 0$$

The stresses can therefore be deduced directly from the strains with a multiplier equal to the shear modulus μ.

In cylindrical coordinates:

$$\sigma_{rr} = (\lambda + 2\mu)\,\varepsilon_{rr} + \lambda\varepsilon_{\theta\theta} + \lambda\varepsilon_{zz}$$

$$\sigma_{\theta\theta} = \lambda\varepsilon_{rr} + (\lambda + 2\mu)\,\varepsilon_{\theta\theta} + \lambda\varepsilon_{zz}$$

$$\sigma_{zz} = \lambda\varepsilon_{rr} + \lambda\varepsilon_{\theta\theta} + ((\lambda + 2\,\mu)\,\varepsilon_{zz}$$

$$\sigma_{rz} = \mu\varepsilon_{rz},\ \sigma_{\theta z} = \mu\varepsilon_{\theta z},\ \sigma_{r\theta} = \mu\varepsilon_{r\theta}$$

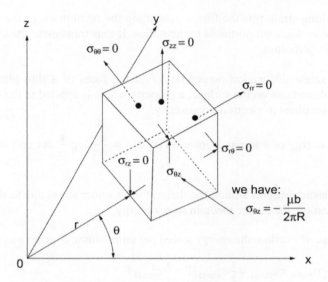

Figure 1.26. *Stresses (in the screw case)*

$$\sigma_{\theta z} = -\frac{\mu b}{2\pi r}, \ \sigma_{rr} = \sigma_{\theta\theta} = \sigma_{zz} = \sigma_{r\theta} = \sigma_{rz} = 0$$

The stresses are thus reduced to two shearing operations for $\sigma_{\theta z}$ and $\sigma_{z\theta}$, with a respective size of $\mu b / 2\pi r$ (see Figure 1.26).

NOTE.– These stresses do not exert force on the side walls of the tube.

At the end of the dislocations, these stresses exert a torsional force, for which the torque around the axis is calculated as follows:

$$dF = \sigma_{z\theta} ds = \sigma_{z\theta} \, 2\pi \, r \, dr$$

$$d\Gamma = \sigma_{z\theta} \, 2\pi \, r^2 \, dr$$

$$\Gamma = \int_{r_0}^{R} d\Gamma = \frac{\mu b}{2} (R^2 - r_0^2)$$

with $r_0 \simeq 5b$, and R for a small $\sigma_{z\theta}$.

The deformation under the action of the torque is helical.

The resulting strain tilts the forces generating the revolutions of the cylinder on the z axis by an angle proportional to the radius. It thus transforms the cylinder into a revolving hyperboloid.

Thus, a screw dislocation perpendicular to the faces of a thin plate produces appreciable distortions on the surface, a property which is applied in the observation of screw dislocations in electron microscopy.

The line energy of a screw dislocation is: $W_L \simeq \dfrac{\mu b^2}{4\pi} \log \dfrac{R}{r_0}$ per unit length of the dislocation.

As the elastic medium around the dislocation is under stress due to the presence of this dislocation, the elastic medium stores energy.

In the case of traction, the energy stored per unit volume is given as:

$$W = \frac{1}{2}\sigma_{maxi} \times \varepsilon_{maxi} = \frac{1}{2}\sigma_{maxi}^2/E = \frac{1}{2}\varepsilon_{maxi}^2 E$$

For a generalized field of stresses, the energy stored per unit volume is given as:

$$W = \frac{1}{2}\left(\sigma_{xx}\varepsilon_{xx} + \sigma_{yy}\varepsilon_{yy} + \sigma_{zz}\varepsilon_{zz} + \sigma_{xy}\varepsilon_{xy} + \sigma_{xz}\varepsilon_{xz} + \sigma_{yz}\varepsilon_{yz}\right)$$

This can be written as a function of strains and/or stresses alone, that is:

$$W = \frac{1}{2}(\lambda + 2\mu)\left(\varepsilon_{xx} + \varepsilon_{yy} + \varepsilon_{zz}\right)^2 + \frac{1}{2}\mu(\varepsilon_{xy}^2 + \varepsilon_{xz}^2 + \varepsilon_{yz}^2$$
$$- 4\varepsilon_{yy}\varepsilon_{zz} - 4\varepsilon_{xx}\varepsilon_{zz} - 4\varepsilon_{xx}\varepsilon_{yy})$$

$$W = \frac{1}{2\mu}\left\{\frac{\lambda + \mu}{3\lambda + 2\mu}(\sigma_{xx}^2 + \sigma_{yy}^2 + \sigma_{zz}^2) + \right.$$
$$+ (\sigma_{xy}^2 + \sigma_{xz}^2 + \sigma_{yz}^2) -$$
$$\left. - \frac{\lambda}{3\lambda + 2\mu}(\sigma_{xx}\sigma_{zz} + \sigma_{xx}\sigma_{yy} + \sigma_{yy}\sigma_{zz})\right\}$$

λ, μ: Lamé coefficients.

For a screw dislocation, we have observed that:

$$\sigma_{xx} = \sigma_{yy} = \sigma_{zz} = \sigma_{xy} = 0$$

The expression of w is simplified to:

$$W_{vis} = \frac{1}{2\mu}(\sigma_{xz}^2 + \sigma_{yz}^2)$$

While:

$$\sigma_{xz} = \frac{\mu b}{2\pi}\frac{y}{x^2 + y^2} \text{ and: } \sigma_{yz} = -\frac{\mu b}{2\pi}\frac{x}{x^2 + y^2}$$

We obtain:

$$W = \frac{1}{2\mu}\left(\frac{\mu b}{2\pi}\right)^2 \frac{1}{x^2 + y^2} = \frac{1}{2\mu}\left(\frac{\mu b}{2\pi}\right)^2 \frac{1}{r^2}$$

And the internal energy per unit length of the dislocation line is given by:

$$W_{Lvis} = \frac{1}{2\mu}\left(\frac{\mu b}{2\pi}\right)^2 \int_{r0}^{R} \frac{2\pi r\, dr}{r^2} \approx \frac{\mu b^2}{4\pi}\log\frac{R}{5b}$$

For a dislocation core $r_0 \approx 5b$, the energy of the core would be in the order of 10% of W_L for a stress in the core of the order of $\mu/30$. In general, the value of R varies from 10^5 b to 10^8 b.

1.3.2.3.4. Stresses, displacement field and strains of a stationary edge dislocation

For the edge dislocation, the dislocation line is perpendicular to its Burgers vector. Let 0z be the axis parallel to the dislocation.

To check the equilibrium equation:

$$(1 - 2v)\,\Delta\vec{u} + \text{grad div } \vec{u} = 0$$

We may consider a local displacement:

$$U_r = \frac{b\theta}{2\pi}$$

But this displacement does not confirm the general equation.

Nabarro (1952) proposed a series of corrective terms to satisfy the general equation.

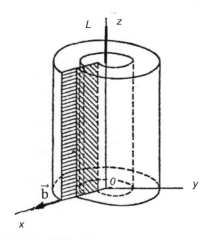

Figure 1.27. *Displacement at the cut*

– Displacement:

$$\vec{U} \begin{cases} U_z = 0 \\ U_r = \dfrac{b}{2\pi}\left(\theta + \dfrac{\sin 2\theta}{4(1-v)}\right) \\ U_\theta = -\dfrac{b}{2\pi}[\dfrac{1-2v}{2(1-v)}\log r + \dfrac{\cos 2\theta}{4(1-v)}] \end{cases}$$

[1.3]

– Stresses in cylindrical coordinates:

$$\sigma_{rr} = \sigma_{\theta\theta} = \dfrac{\mu b}{2\pi(1-v)}\dfrac{\sin\theta}{r}$$

$$\sigma_{zz} = \dfrac{\mu v b}{\pi(1-v)}\dfrac{\sin\theta}{r}$$

$$\sigma_{r\theta} = -\dfrac{\mu b}{2\pi(1-v)}\dfrac{\cos\theta}{r}$$

[1.4]

We note the existence of stress σ_{zz} which acts parallel to the line L of the edge dislocation, despite the lack of displacement in this direction.

– The elastic displacements of the equations [1.3] give rise to the deformations given in x, y, z coordinates:

$$\varepsilon_{xx} = \frac{by}{2\pi} \frac{\mu y^2 + (2\lambda + 3\mu)x^2}{(\lambda + 2\mu)(x^2 + y^2)^2}$$

$$\varepsilon_{yy} = -\frac{by}{2\pi} \frac{(2\lambda + \mu)x^2 - \mu y^2}{(\lambda + 2\mu)(x^2 + y^2)^2}$$

$$\varepsilon_{xy} = -\frac{b}{2\pi(1-v)} \frac{x(x^2 - y^2)}{(x^2 + y^2)^2}$$

$$\varepsilon_{zz} = \varepsilon_{xz} = \varepsilon_{yz} = 0$$

[1.5]

– The stresses calculated from these strains are:

$$\sigma_{xx} = \frac{\mu b}{2\pi(1-v)} \frac{y(3x^2 + y^2)}{(x^2 + y^2)^2}$$

$$\sigma_{yy} = -\frac{\mu b}{2\pi(1-v)} \frac{y(x^2 - y^2)}{(x^2 + y^2)^2}$$

$$\sigma_{zz} = v(\sigma_{xx} + \sigma_{yy}) = \frac{\mu v b y}{\pi(1-v)(x^2 + y^2)}$$

$$\sigma_{xy} = -\frac{\mu b}{2\pi(1-v)} \frac{x(x^2 - y^2)}{(x^2 + y^2)^2}$$

$$\sigma_{xz} = \sigma_{yz} = 0$$

[1.6]

– Representation of stresses: for $\frac{\mu b}{2\pi(1-v)r} = 1$, a representation of the stresses [1.6] is shown (see Figure 1.28) according to θ.

The calculation of the energy W_L per unit length of the edge dislocation line is equal to:

$$W_{Lcoin} = \frac{\mu b^2}{4\pi K} \log \frac{R}{r_0}$$

with:

$$K = 1 - v$$

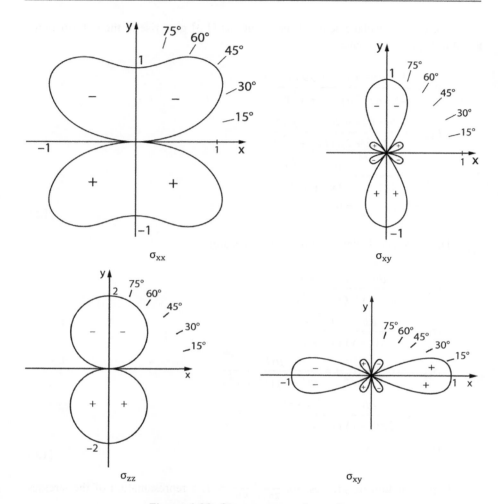

Figure 1.28. *Stresses around an edge dislocation (Kovacs and Zsoldos 1973)*

This energy W_{Lcoin} is greater than that of a dislocation screw by a factor of $1/(1 - v) \approx 3/2$.

In the case of aluminum with $E = 69 \times 10^3$ MPa: $v = 0.34$, we obtain:

$$\mu = \frac{E}{2(1 + v)} \approx 26 \times 10^3 \text{ MPa}$$

$b \approx 2.9 \times 10^{-10}$m and, for R = 10^5b, the energy of a screw dislocation line is in the order of 1.8×10^{-4} ergs/cm, and 2.7×10^{-4} ergs/cm for an edge dislocation:

(1 erg = 10^{-7}J)

The signs in the graphs indicate the tensions (+) and the compressions (−).

The shear values σ_{xy} are maximum for x = 0.

$$\frac{\mu b}{2\pi (1 - \nu)r}$$

These are equal to 1 for the figures (see Kovács and Zsoldos 1973).

The Burgers vector $\vec{b}$ of the mixed dislocation forms an arbitrary angle with the dislocation line L; we will use the z axis on the line L and an x axis along the slip plane of the dislocation (Figure 1.30).

We obtain: $b_x\vec{\imath}$ is ⊥ à L (edge component of $\vec{b}$).

$b_z\vec{k}$ is the screw component of $\vec{b}$.

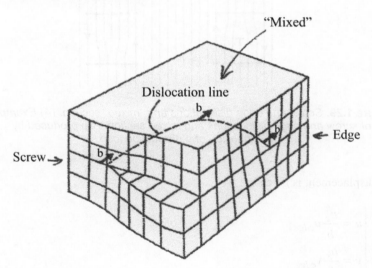

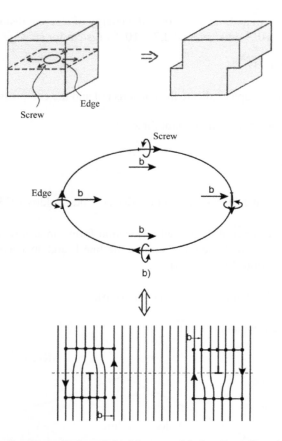

Figure 1.29. *Screw and edge dislocations and "mixed" section. (A) Example of screw and edge dislocations and mixed section. Slip produced by the expansion of a dislocation loop*

The displacement is given by:

$$\vec{u}\begin{cases} u = \dfrac{b_x}{b}u_{edge} \\[2ex] v = \dfrac{b_x}{b}v_{edge} \\[2ex] w = \dfrac{b_z}{b}w_{screw} \end{cases}$$

[1.7]

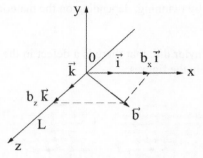

$$\vec{b} = b_x \, \vec{i} + b_z \, \vec{k}$$

The length of the mixed Burgers vector is:

$$|\vec{b}| = (b_x^2 + b_z^2)^{1/2}$$

Figure 1.30. *Mixed Burgers vector*

The difference of the vectors of the displacements representing the final point and the initial point is equal to the Burgers vector $\vec{b}$.

For u_{edge}, v_{edge} and w_{screw}, we can return to the equations previously described and state that the strains caused by the mixed dislocation are the sum of the strains of a pure edge dislocation of the Burgers vector $b_x\vec{i}$ and strains of a pure screw dislocation of the Burgers vector $b_z\vec{k}$. The same applies to stresses.

The energy per unit length of a mixed dislocation is given by:

$$W_L = \frac{\mu b^2}{4\,\pi K} \log \frac{R}{r_o}$$

with:

$$\frac{1}{K} = \cos^2 \varphi + \frac{\sin^2 \varphi}{1 - \nu}$$

where φ is the angle formed by the dislocation and its Burgers vector: for a screw $\varphi = 0$; for an edge $\varphi = \frac{\pi}{2}$; $r_o \simeq 5b$.

1.3.3. *Crystallographic slips*

On a microscopic scale, metals deform anisotropically. The relative displacements of the atoms under the effect of external stresses mainly occur through slipping and/or by twinning, depending on the material and the experimental conditions of the strain.

Let us imagine, as Taylor did, that there is a defect in the network at the level of the slip plane, as shown in the figures.

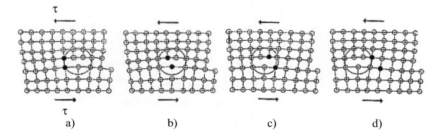

Figure 1.31. *Movement of a dislocation in a crystal*

The displacement of this defect known as dislocation, progressing one by one, results in the formation of a superficial step of height B. If such a defect stretches across the entire single crystal, this then generates a slip of amplitude B. The energy used in moving the defect simply represents the amount needed to switch the atomic bonds of a single row. Therefore, this energy and the corresponding stress level are considerably reduced.

Mott, another English physicist, provided a telling image of this by noticing that it is very easy to move a carpet placed on top of another carpet by spreading a fold in the carpet.

In the general case, the plastic strain is therefore intimately linked to the movement of the dislocations, which we will need to study on the basis of the materials, conditions and types of impacts.

The slip is characterized by a shearing of the crystal lattice along a plane and following a direction, which causes the translation of an integer number of interatomic distances from the upper part relative to the lower part. The slip systems differ between the cubic face-centered (CFC), hexagonal compact (HC) and cubic centered (CC) crystal structures (Figure 1.33).

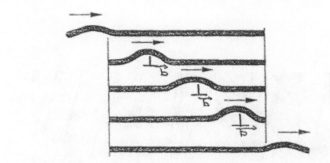

Figure 1.32. *Mott's carpet*

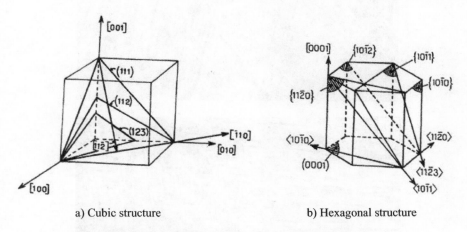

a) Cubic structure   b) Hexagonal structure

Figure 1.33. *Main crystallographic planes involved in the slip*

Any displacement of a dislocation line generally requires either an addition or a removal of material. This non-conservative movement is called the climb. However, a displacement of the line within the plane defined by the line and the Burgers vector $\vec{b}$ is conservative: this is known as a slip.

REMARK.–

– In the case of a rectilinear dislocation, the slip surface contains L and $\vec{b}$.

– A screw dislocation can slip in any plane, since L and $\vec{b}$ are equal.

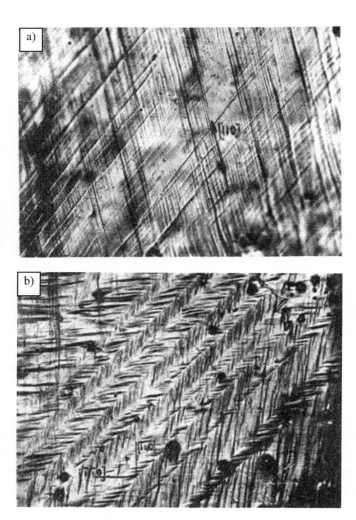

Figure 1.34. *Optical micrographs (G = 250) of slip lines, in the case of copper mono-crystals subjected to stress by electromagnetic pulses (10 MPa in ~ 15 μs): in (a) crystal [111], double slip (T = 78 K); in (b) crystal [100], slip bands, stress at 5 K (Leroy 1972)*

1.3.3.1. *Crystal slip systems*

Crystal	Lattice type	1) Glide direction	2) Glide plane	Remarks
Cu, Ag, Au, Ni, α-CuZn	Al-type fcc	$\langle 110 \rangle$	$\{111\}$	
Al	Al-type fcc	$\langle 110 \rangle$ $\langle 110 \rangle$	$\{111\}$ $\{110\}$	above 450 °C
NaCl, KCl, KJ, KBr	B1-type fcc	$\langle 110 \rangle$	$\{110\}$	
PbTe	B1-type fcc	$\langle 100 \rangle$	$\{110\}$	
Si, Ge	A4-type fcc	$\langle 110 \rangle$	$\{111\}$	
α-Fe	A2-type bcc	$\langle 111 \rangle$ $\langle 111 \rangle$ $\langle 111 \rangle$	$\{110\}$ $\{112\}$ $\{123\}$	
α-Fe + 4% Si, Mo, Nb	A2-type bcc	$\langle 111 \rangle$	$\{110\}$	
AuZn, AuCd, MgTl, NH₄Cl, TlCl, NH₄Br	B2-type cubic	$\langle 100 \rangle$	$\{100\}$	
AgMg	B2-type cubic	$\langle 111 \rangle$	$\{321\}$	
β′-CuZn	B2-type cubic	$\langle 111 \rangle$	$\{110\}$	
Cd, Zn ZnCd	A3-type hcp c/a = 1.85	$\langle 2\bar{1}\bar{1}0 \rangle$ $\langle 2\bar{1}\bar{1}0 \rangle$ $\langle 2\bar{1}\bar{1}3 \rangle$	$\{0001\}$ $\{01\bar{1}1\}$ $\{\bar{2}112\}$	
Mg	A3-type hcp c/a = 1.623	$\langle 2\bar{1}\bar{1}0 \rangle$ $\langle 2\bar{1}\bar{1}0 \rangle$ $\langle 2\bar{1}\bar{1}0 \rangle$	$\{0001\}$ $\{01\bar{1}1\}$ $\{01\bar{1}0\}$	above 225 °C below room temp.
Be	A3-type hcp c/a = 1.568	$\langle 2\bar{1}\bar{1}0 \rangle$ $\langle 2\bar{1}\bar{1}0 \rangle$	$\{0001\}$ $\{01\bar{1}0\}$	
Ti	A3-type hcp c/a = 1.587	$\langle 2\bar{1}\bar{1}0 \rangle$ $\langle 2\bar{1}\bar{1}0 \rangle$ $\langle 2\bar{1}\bar{1}0 \rangle$	$\{01\bar{1}0\}$ $\{01\bar{1}1\}$ $\{0001\}$	

Table 1.5. *Slip systems: (1) slip directions; (2) slip planes (Seeger 1958)*

1.3.3.1.1. Braking force

Causing a dislocation to slip from one atomic position to another requires a certain amount of energy. The network essentially exerts a braking force on the movement of the dislocations. Peierls and Nabarro have shown that this force is strong when leaving dense planes, while it is weak in dense planes. The movement of the dislocations therefore occurs in dense planes; the dislocations tend to follow the dense rows (see Tables 1.5 and 1.6).

The crystals are classified into two categories according to their Peierls and Nabarro strengths:

– small braking forces (H, CFC, CC);

– large braking forces (diamond structure, Ge, Si).

These forces and their influences will be listed.

1.3.3.2. *Example of dislocation displacements: visualization of dislocations by "attack figures" taking the case of lithium fluoride[1]*

1.3.3.2.1. Preparation

Why was this choice made?

Methods are known for preparing certain alkali metal halides, such as LiF, with a very high degree of purity and very few structural defects. Using cleavage, a much more effective method than cutting, excellent samples can be obtained, and the nature of the material lends itself well to the easy detection of the presence of dislocations. These halides are thus well suited for the experimental study of the properties of dislocations. In particular, the development of techniques for observing dislocations in lithium fluoride has made it possible to make significant progress in understanding the phenomenon.

Methods exist for preparing very pure lithium fluoride monocrystals and faces strictly parallel to the {100} planes are obtained by cleavage. Two particular experimental techniques can be used, which when combined form an effective method of examining the processes of strain. These include both the "attack figure" technique to reveal the positions of the dislocations on the surface of the crystal and the optical technique of observation of the birefringence, which provides the distribution of the strains (Newey and Davidge 1965).

---

1 Excerpt from practical works, Imperial College, Surrey.

The first technique is based on the fact that the energy of a crystalline surface is higher at the emergence point of a dislocation, due to the deformation energy associated with the dislocation.

It is then possible to detect the position of the dislocation by immersing the crystal in a solution in which it is slightly soluble. With suitable solutions, the region of the crystal that is very close to the dislocation dissolves faster than the surrounding surface with greater regularity. Attack patterns occur on the surface; the size and shape of these patterns depend on the solution used, the attack time, the nature of the structure and the crystallographic orientation of the surface.

The attack figures (Figure 1.35) are obtained in lithium fluoride, using hydrogen peroxide at 20 volumes. The sample is placed in a cup and is covered with a freshly prepared solution and left to sit for about 20 min. It is rinsed carefully in water, with ethyl alcohol, and finally with ether.

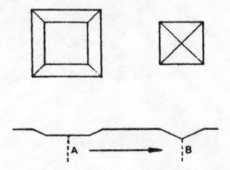

Figure 1.35. *Detection of the movement of a dislocation by double attack: the dislocation was in A during the first attack and in B during the second attack. (Newey and Davidge 1965)*

The optical method is based on the fact that some parent crystalline substances become birefringent when elastic strains occur within them. Under suitable lighting conditions with polarized light, these strains are made visible through variations in contrast. In this way, it is possible to detect the regions of the crystal which have slipped under the action of the elastic strains due to the dislocations.

Under the action of the compression of the crystal, two groups of stresses can be identified:

– First, an axial compressive stress with the birefringence associated with it, and which is canceled when the compression is removed.

– For a compression level that is sufficient to allow plastic strain, additional elastic stresses are located near the active slip planes. The primary stresses appear in the directions [110] and perpendicular to the active slip planes, thus presenting a birefringence which can be observed in the vicinity of these planes.

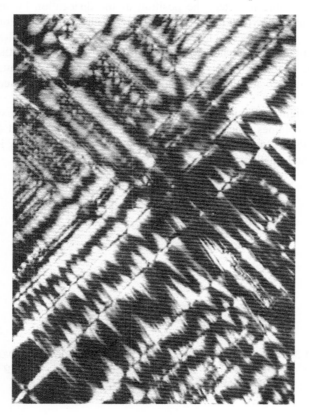

Figure 1.36. *When no stress is present, the lithium fluoride is isotropic, thus the crystal appears black for all orientations between crossed polarizers and analyzers. In the image, it can be found that in a stressed state and for a suitable position of both the crystal as well as the polarizer–analyzer assembly, the strained regions of the crystal are highlighted (Newey and Davidge 1965)*

1.3.3.2.2. Types of lithium fluoride slips

The lithium fluoride slips along planes of the {110} type (diagonal planes); the six planes of this family are represented in Figure 1.37. The related slip vector is of the type <110>; each vector occurs within one of the planes {110}. The crystal therefore has six slip systems. If the crystal is compressed in one direction [001], it can slip along four of the possible slip planes: $(0\bar{1}1)$, (011), $(\bar{1}01)$ and (101) (Figure 1.37). These four planes form the same angle with the direction of the stress. They thus undergo equal shearing forces. The planes $(1\bar{1}0)$ and (110) are parallel to the direction of the stress, so the component of the corresponding shear stress is zero.

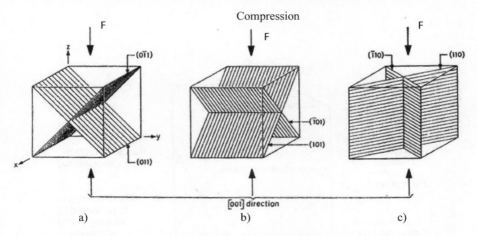

Figure 1.37. *The six slip planes of lithium fluoride; in (a) and (b), four possible planes of slip through compression due to the force F according to [001]*

1.3.3.2.3. Compression of the crystal

Figure 1.38 shows the observations made by leading figures and by birefringence, which reveal that the slip planes are of the type (110). Since we can only see in front of the slip steps in the directions <100>, these rows <100> represent screw dislocations; under the action of the applied force, the loop extends and the step develops toward D (Figure 1.39).

The rows of attack figures (face B) in the directions <110> of the crystal correspond to dislocations.

It is understood that the slip occurs with a Burgers vector of the type <110>, considering the crystal structure of the LiF (Figure 1.40). In a given crystal structure, the Burgers vector of the dislocations of the slip, as a general rule, is the smallest vector joining the equivalent ion sites. For example, Figure 1.40 shows three such vectors along the directions <100>, <110> and <111>, and they have for lengths a, $a/\sqrt{2}$, $a\sqrt{3}$, respectively, where a is the edge of the elementary mesh. Since the energy of a dislocation is proportional to the square of its Burgers vector, the energies of the dislocations listed above are proportional to a^2, $a^2/2$ and $3a^2$, respectively. Note that, for these directions <100> and <111>, the ions are alternatively positive and negative. The corresponding vectors are therefore equal to twice the difference between adjacent ions. The dislocations of lower energy thus correspond to the vector <110>. These are dislocations that will mainly participate in the strain, which is perfectly in keeping with the observed results.

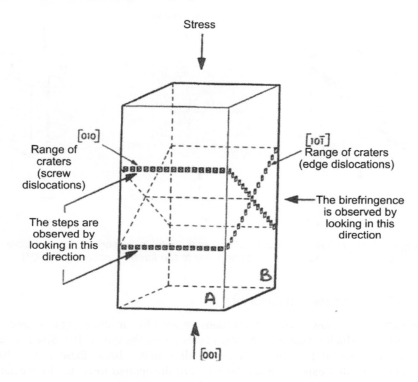

Figure 1.38. *A series of observations on the strained crystal (Newey and Davidge 1965)*

NOTE.– The directions <110> are common to many crystal planes, such as {100}, {110} and {211}. In many crystals, the Burgers vector is located within the densest atomic planes; for lithium fluoride, these are the {100} planes, but observations show that the slip planes are of the {110} type. The reasons why it is the lower density {110} planes at work here are quite complex and cannot be explained within the framework of this presentation. However, it can be said that the main factor of the phenomenon is related to the repellent forces between ions of the same charge that are led to be brought together when the dislocation moves. In LiF, this effect is less for dislocations moving in {110} planes than for those moving in {100} planes.

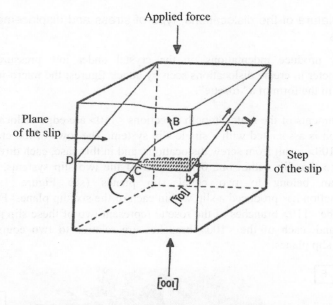

Figure 1.39. *ABC dislocation quarter loop along the slip plane [101]*

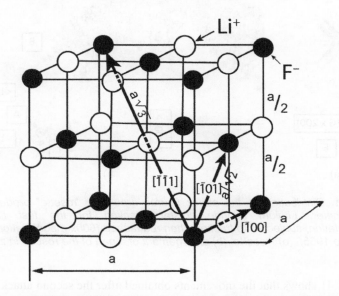

Figure 1.40. *Examples of Burgers vectors <100>, <110> and <111>*
in the structure of lithium fluoride. The active vector is of the type <110>

1.3.3.2.4. Nature of the dislocations, type of stress and displacement of the dislocations

We may produce indentations on the crystal under low pressure using a micro-durometer to create dislocations seen by attack figures; the micro-indentation is said to be in the form of a "rosette".

The alignments of the crater in both directions <110> are edge dislocations, and each direction is associated with a single slip system. The rows of the crater in the directions <100> result from screw dislocations, and in this case, each direction may correspond to the slip within one or the other of the two slip systems, since this direction can belong to two equal slip planes (see Figure 1.37). The micro-indentation has produced a slip within each of the six slip planes. Figure 1.42 shows that the <110> branches of the rosette represent two of these slip planes. On the other hand, each of the <100> branches are relative to two equivalent but non-parallel slip planes.

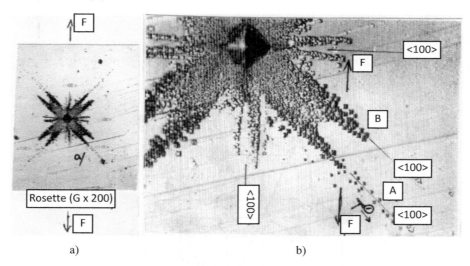

a)   b)

Figure 1.41. *Distribution of the attack figures around a "rosette" produced by a micro-durometer. Lithium fluoride crystal attacked for the first time after micro-indentation, subjected to traction, then reattached (600x magnification) (Newey and Davidge 1965). (a) Rosette; (b) enlargement of a part of the rosette in zone (a)*

Figure 1.41 shows that the movements obtained after the second attack are made in four of the eight branches of the rosette which contain the dislocations (<110> branches).

The direction of the movements of the dislocations will be determined by their sign – which is to say, by the relative position of their half-planes – and the translation of the region containing the half-plane occurs in the direction of the movement of the dislocation (see Figure 1.42 in the case of traction).

The <100> branches contain screw dislocations, which move more slowly than the corners with the same stress. Thus, we obtain lower displacements.

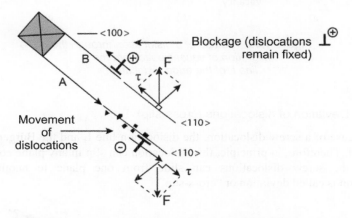

Figure 1.42. *The movements of the edge dislocations of opposite signs within a branch of the rosette acted upon by the tension force F, τ is the shear due to F along the slip plane*

NOTE.– In the case of compression, the movement of edge dislocations is in the opposite direction of that shown in Figure 1.42.

1.3.3.3. *The climb of a dislocation*

If the network contains vacant sites (vacancies), the dislocation can move perpendicularly to the slip plane through the diffusion of the vacancies toward the dislocation line. This climbing movement is illustrated in Figure 1.43.

The contribution may also be due to interstitial atoms along the dislocation line. These vacancy or interstitial "climbs" occur especially at high temperatures.

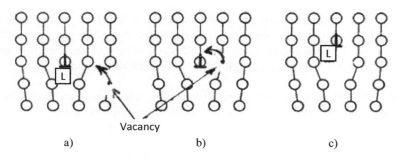

Vacancy

a)   b)   c)

Figure 1.43. *The rise of a dislocation through the migration of vacancies toward the line L of the edge dislocation*

1.3.3.3.1. Deviation of dislocations (cross-slip)

In the case of a screw dislocation, the dislocation line L and the Burgers vector are parallel. Therefore, in principle, this dislocation can slip in any plane containing it. Thus, the screw dislocations can slide from one plane to another; this phenomenon is called deviation or "cross-slip".

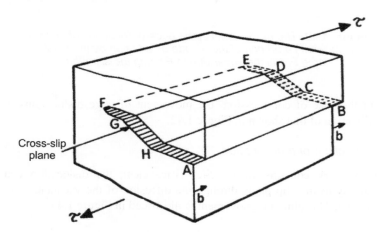

Figure 1.44. *Cross-slip of a screw dislocation: a screw dislocation created in AB has slipped along the plane ABCH, then in the parallel plane DEFG via the plane CDGH. This CDGH plane is called the cross-slip plane*

Thus, considerable changes occur in the position of the dislocations. These involve the annihilation of numerous dislocations (with the dislocations of opposite signs being canceled when they coincide) and the formation of stable arrangements of dislocations (such as those encountered in the sub-grain boundaries). Such processes allow strain to be reduced and the crystal to be restored. The movements of the dislocations during the recovery will be described later.

1.3.4. *Twinning*

For a slip, the passage of a dislocation in the crystal leaves no trace, and we are given a shear equal to one period of the network, which is reconstituted without disturbance behind the dislocation.

By contrast, twinning is a mode of deformation which is accompanied by a crystalline modification in volume. It can be described as a homogeneous shearing of a part of the crystal located between two habit planes or twin planes. The shearing between two neighboring planes is less than a network vector. The twinned part has tilted overall and has come into a position symmetrical to that of the matrix with respect to the twin plane (Figure 1.45). The twins form quickly and spread at a very high speed in the crystals.

This is a particularly important strain mechanism since most metals and alloys lend themselves very easily to twinning with high strain speeds.

The ease with which the twinning occurs depends on various factors: pressure, crystallographic orientation, stacking fault energy, the impulse duration, microstructure, and initial grain size. Twinning may become the preferred method of strain.

Figure 1.46 shows the influence of temperature on the critical stress necessary for twinning, and the slip of dislocations at low and high deformation rates.

In Figure 1.46, the temperature T_M below which the flow of the material occurs through twinning, and which corresponds to the intersection of the curves representing the critical stresses necessary for the twinning and slip of dislocations, increases with the speed of the strain.

The nucleation of a mechanical twin can be initiated from heterogeneous circumstances, such as the particular arrangement of dislocations, thus creating a stacking fault.

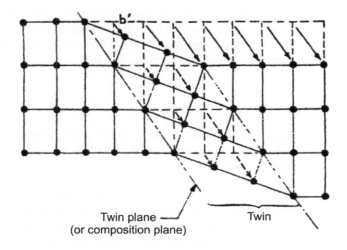

Figure 1.45. *Mechanical twin (Adda et al. 1979)*

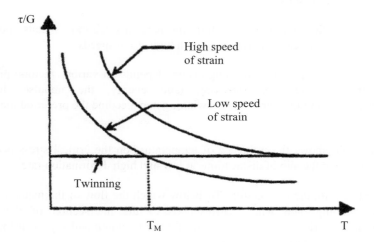

Figure 1.46. *Twinning zone*

A twin can be created by the movement of partial dislocations. Let us suppose, for example, that a Shockley dislocation with a Burgers vector (a/6) [11$\overline{2}$] moves through all of the compact planes (111) contained in a portion of a CFC crystal. The stacking of this portion, initially of the ABCABC type, would become CBACBA.

We will assume in this discussion that the partial dislocations are related to simple stacking defects. Figure 1.47 shows the dislocations in the process of changing the order of the stack; the modified region is called the twinned region.

The orientation of the new crystal, the twin, is the orientation of the initial crystal reflected by a mirror located in the plane of the boundary.

The orientation of the twin with respect to the untwinned crystal is identical to the relative orientation of the two components of a bicrystal constructed by joining the plane PQR of one of the cubes of Figure 1.47 to the plane P'Q'R' of the other cube (and removing the additional plane of atoms), such that P' joins to P, Q' to Q, and R' to R.

The result of this grafting is that the PQR plane, or P'Q'R', is a reflection plane.

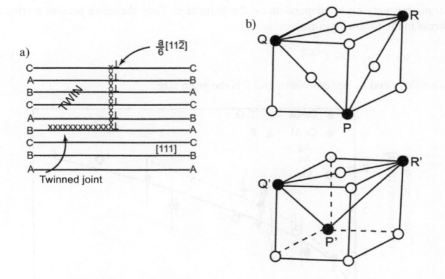

Figure 1.47. *Dislocations that change the stacking order. (a) Stacking order of a twin. (b) The twin is constructed by joining P and P', Q and Q', R and R' and by removing the additional atomic plane*

Friedel (1964) proposes a critical twinning stress resulting from a localized arrangement of dislocations in the following form:

$$\tau_c = \frac{\gamma_{fe}}{nb}$$
[1.8]

where γ_{fe} is the stacking fault energy of the material, b is the Burgers vector of the dislocations that generate the microtwinnings and n is a stress concentration factor $(1 \leq n \leq 3)$.

The stress concentration factor makes it possible to explain the formation of twins in certain areas of the crystal and the existence of high densities of dislocations in other areas.

In steels, Zerilli and Armstrong (1987) have found that the critical stress at which twinning appears is relatively independent of the temperature and the speed of strain, but very highly dependent on the grain size. They therefore present a critical stress in the following form:

$$\sigma_T = \sigma_{T0} + k_T \, \ell^{-1/2}$$
[1.9]

where σ_{T0} and k_T are constants, and ℓ is the grain size.

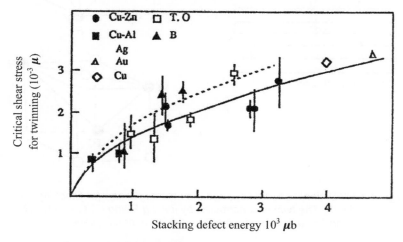

Figure 1.48. *Variation of the theoretical stress of twinning as a function of the stacking fault energy (dotted curve): results for Cu Zn, Cu Al, Ag and Au alloys on copper monocrystals (solid-line curve) (François et al. 1991/1992, p. 276)*

System	Substance	Slip		Twin		
		Plane	Direction	Plane	Direction	Plane conjugate
CFC (Cubic, faces centered)	Metals and alloys	(111)	[10$\bar{1}$]	(111)	[11$\bar{2}$]	(11$\bar{1}$)
CD (Cubic, diamond type)	C, Si, Ge	(111)	[10$\bar{1}$]	(111)	[11$\bar{2}$]	(11$\bar{1}$)
CC (Cubic, centered)	Feα, CuZn, Mo	(1$\bar{1}$0)	[11$\bar{1}$]	(112)	[11$\bar{1}$]	(11$\bar{2}$)
	Feα, Mo, W, Na	(112)	[11$\bar{1}$]	(112)	[11$\bar{1}$]	[11$\bar{2}$]
	Feα, K	(123)	[11$\bar{1}$]	(112)	[11$\bar{1}$]	[11$\bar{2}$]
HC (Hexagonal, close-packed)	Be, Ti, Mg	(0001)	[2$\bar{1}\bar{1}$0]	(10$\bar{1}$2)	[$\bar{1}$011]	(10$\bar{1}$2)
	Zn, Cd	(0001)	[2$\bar{1}\bar{1}$0]	(10$\bar{1}$2)	[10$\bar{1}\bar{1}$]	(10$\bar{1}$2)
O (Orthorhombic)	Uα	(010)	[100]	(130) (112) (121)	[3$\bar{1}$0] [312]	($\bar{1}$10) (112)
R (Rhombohedral)	As, Sb, Bi	(111)	[100] and [10$\bar{1}$]	(110)	[00$\bar{1}$]	(001)
	Hg	(100)	–	(110)	[00$\bar{1}$]	(001)
T (Tetragonal)	Snβ, In	(110), (100), (010), (10$\bar{1}$) and (12$\bar{1}$)	[001] [101]	(301) (101)	[$\bar{1}$03] [101]	($\bar{1}$01) ($\bar{1}$01)

Table 1.6. *Typical slip and twinning elements for certain crystalline systems*

1.3.5. *Force on the dislocation*

A shear stress is applied τ on an element of material (rectangular parallelepiped L, ℓ, h). For a displacement of the two half-blocks of a value b (Figure 1.49), the work W performed is:

$$W = F \times b = \tau A \times b$$

With:

$$A = L \times \ell$$

where F = shear force in the surface slip plane A.

If the displacement of a dislocation is dL for the blocks, we obtain b x dL/L with a work dW equal to:

$$dW_{(\tau)} = \tau A \times dL = \tau \ell b\, dL \qquad [1.10]$$

For a displacement of the dislocation line by a value dL under the action of a force F, the work of this force is:

$$dW_{(F)} = FdL \qquad [1.11]$$

Setting the work values in [1.10] and [1.11] equal makes it possible to obtain the force F exerted along the length of the line ℓ of the dislocation, or:

$$dW = \tau \ell b\, dL = FdL \qquad [1.12]$$

$F = \tau \ell b$, per unit of dislocation length: $F = \tau b$, hence a possible displacement of the dislocation line by shearing action τ.

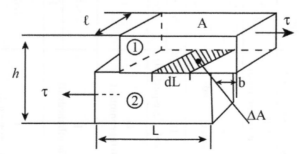

Figure 1.49. *Slip of blocks 1 and 2 over a quantity b*

Obstacles and Mechanisms
of Crossings

Recall that under the action of a shear stress τ, the force exerted on a dislocation of a Burgers vector $\vec{b}$ per unit length of the dislocation is equal to:

$$F = \tau b \qquad\qquad [2.1]$$

The deformation is a slip with the value of ΔL with a slip angle of $\Delta\gamma$ such that:

$$\Delta\gamma = b\frac{\Delta L}{L} / h \qquad\qquad [2.2]$$

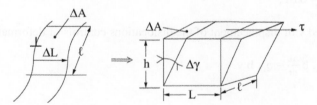

Figure 2.1. *Strain from shear*

We introduce the volume V of the block and the area ΔA:

$$\Delta\gamma = \frac{b\,\Delta L}{hL}\cdot\frac{\ell}{\ell} = b\frac{\Delta L \times \ell}{V} = b\frac{\Delta A}{V} \qquad\qquad [2.3]$$

If the block of material of volume V contains N dislocations of the same Burger vector b, we define the density of dislocations ϱ_D as the total length of the dislocations per unit volume, or:

$$\varrho_D = N\ell/V \tag{2.4}$$

Considering that these dislocations move at an average distance of ΔL, the formulas [2.3] and [2.4] give an overall shear strain equal to:

$$\gamma = Nb \frac{\Delta A}{V} = \varrho_D \frac{V}{\ell} b \frac{\Delta A}{V}$$

$$\gamma = \varrho_D b \frac{\Delta A}{\ell} = \varrho_D b \Delta L$$

We obtain:

$$\gamma = \varrho_D b \Delta L \tag{2.5}$$

with: $\varrho_D = Do\ell/V$ moved from ΔL.

In reality, only a fraction of the dislocations are mobile, with their density as ϱ_m, and [2.5] becomes $\gamma = \varrho_m b\Delta L$; from this, the shear rate can be deduced as:

$$\dot{\gamma} = b \frac{d(\varrho_m . \Delta L)}{dt} \tag{2.6}$$

From [2.6], we obtain two limit cases:

– the production speed of mobile dislocations controls the strain speed:

$$\dot{\gamma} = \frac{d\varrho_m}{dt} b.\Delta L;$$

– the speed v of displacement of the dislocations controls of deformation:

$$\dot{\gamma} = \varrho_m b \frac{\Delta L}{dt} = \varrho_m b v$$

2.1. Obstacles

Certain obstacles can oppose the movement of dislocations and control the deformation speeds; the most difficult obstacle to pass primarily controls $\dot{\gamma}$:

– discrete obstacles, whether preexisting as the precipitates or the atoms of the solid solution, or induced by deformation, as with the stress on the "trees" in the forest of dislocations, dislocation cells, etc.;

– diffuse obstacles, like network forces, which are overcome by the disengaging of the dislocations.

2.2. Nature and resistance of obstacles

The obstacles that a dislocation encounters as it slips can be grouped into three
main categories (Table 2.1):

Force of obstacles	Energy	Examples
Strong	$\simeq \mu b^3$	Precipitates bypassed by dislocations
Average	0.2 to μb^3	Dislocations of the forest Small sheared precipitates Irradiation defects
Low	$\leq 0.2\ \mu b^3$	Network forces Solid solution

Table 2.1. *Energy for overcoming certain obstacles*

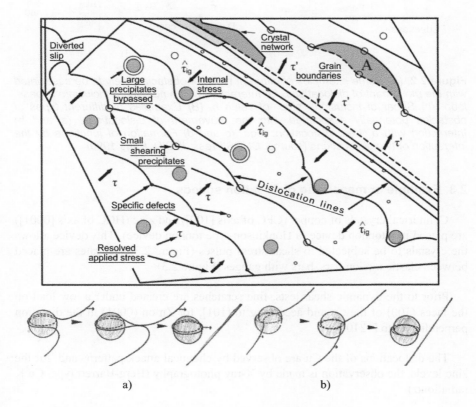

a)   b)

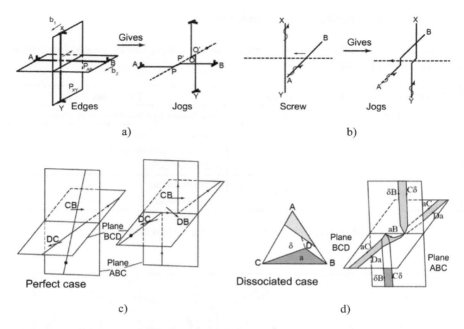

Figure 2.2. *(A) Some examples of different types of obstacles likely to be associated with the movement of dislocations. (a) Interaction with a non-shear inclusion: bypass, loop. (b) Shear of inclusion: "step" of value b. (B) Examples of different types of obstacles potentially associated with the movement of dislocations. (a and b) Interaction with a "tree" dislocation: jogs. (c and d) Formation of junctions by the interaction of two dislocations (source: CNRS Press, IRSID, Paris 1990)*

2.3. Example of measuring dislocation speeds

Cylindrical crystals of copper (CFC of axis [100]) and zinc (HCC of axis [0001]) are placed in a torsion device (a Hopkinson-type torsion device). This device allows the crystals to be subjected to shear stress pulses (Figure 2.3); the rates are placed between elastic torsion bars, bars with gauges (Figure 2.3).

Prior to the dynamic shear tests, fine scratches are created under a low load on the faces (100) of the Cu and according to [101], for Zn on (0001) in the direction perpendicular to [$1\bar{2}10$].

The dislocations of the Cu are observed by chemical attack patterns and, for the zinc levels, the observation is made by X-ray photography (Berg-Barrett type, Cu K radiationα).

Thus the distribution of obstacles more potent at cold strain, and more effectively destroyed, the speed, in the frequency and intensity of a transient the deformation. Consequently, the electromechanically not much values (1-1 × 10⁻⁴ × 10⁻³ diameter) and no time limits, its to like × 10⁻⁴ × 10⁻³ material, this was the 10⁻² AU.)

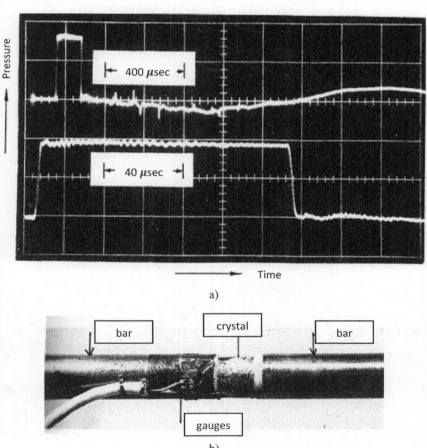

a)

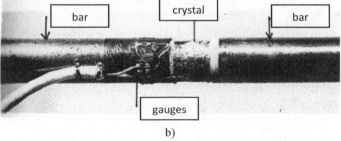

b)

Figure 2.3. *Zinc crystal between elastic monel and titanium bars, and extensometry gauges glued to monel bars. (a) Type of pulse with shear $\tau = C^{te}$ duration $\simeq 120\ \mu s$*

Once the durations of the constant stress pulses τ are known and the displacements obtained, the speeds of the dislocations are deduced as a function of the shear intensities. For example, for copper crystals, we obtain speeds of 10 m/s to 3×10^7 dynes/cm^2 and, for zinc, 6 m/s to 1.8×10^7 dynes/cm^2 (dyne/cm^2 = 10^{-7} MPa).

The experimental curves obtained by Vreeland[1] give:

$$V_{cu} = V_o \frac{\tau}{2.7 \times 10^4}$$

and:

$$V_{zn} = V_o \frac{\tau}{2.94 \times 10^4}$$

with:

$V_o = 1$ cm/s and τ in dynes/cm^2

A comparison between various materials is given below.

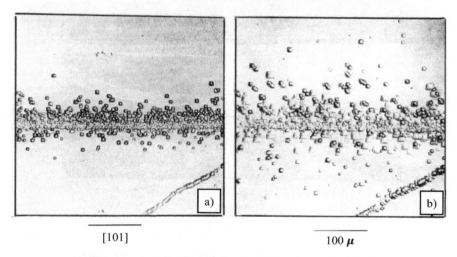

[101]   100 μ

Figure 2.4. *Copper crystal face (100), stripe [101]. (a) Before load is applied. (b) After load is applied, displacements of dislocations*

1 Vreeland, Jr. T. (1968). *Dislocation Dynamics*. McGraw-Hill Book Company, New York.

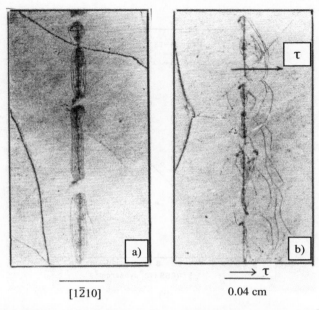

[1$\bar{2}$10]   0.04 cm

Figure 2.5. *Zinc crystal, photographs Rx face (0001) stripe ⊥ [1$\bar{2}$10].*
(a) Before load is applied. (b) After load is applied, displacement τ
dislocations (Vreeland Jr.)

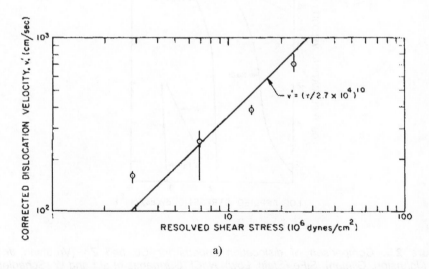

a)

Figure 2.6. *Comparison of dislocation speeds for Cu and Zn (Vreeland Jr.),*
LiF (Johnston, Gilman), SiFe (Stein, Low), NaCl (Gurmanas et al.)
and W (Schadler). (a) Speed of copper dislocations

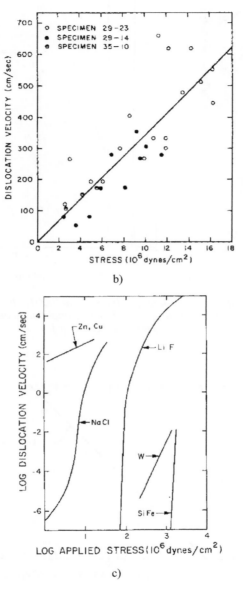

b)

c)

Figure 2.6. *Comparison of dislocation speeds for Cu and Zn (Vreeland Jr.), LiF (Johnston, Gilman), SiFe (Stein, Low), NaCl (Gurmanas et al.) and W (Schadler) (continued). (b) Speed of edge dislocations in zinc. (c) Speeds of dislocations, comparison of different materials*

2.4. Microstructural mechanisms of the deformation rate

Plastic strain is only the macroscopic consequence of displacements on an atomic scale: slip and, to a lesser extent, smearing.

Different mechanisms can exist during the strain, but some predominate and make the influence of the others negligible, which translates into a particular evolution of the flow stress with the temperature and the deforestation rate.

In general, for pure metals and lightly loaded alloys, the evolution of the flow stress with the strain rate can be reflected in the behavior shown in the figure.

This displacement can differ depending on the conditions of deformation speed and temperature, and three types of deformation mechanisms and its speed are generally distinguished:

– the athermal mechanism (area III);

– the thermally activated mechanism (domain I);

– the viscous friction mechanism (area II).

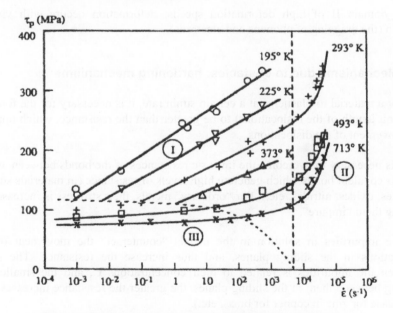

Figure 2.7. *Effect of the strain rate and temperature on the shear yield strength of a mild steel (Campbell et al. 1970)*

The strain of the plastic begins when the force on a dislocation becomes sufficient for it to move, or in other words, once a reduced split in the sliding plane reaches a critical value. This critical split has an athermal component and a thermally activated component.

These athermal processes represent a necessary split τ, which depends little on the speed and temperature. Thermally activated processes result in a split τ, which decreases rapidly when the temperature increases and when the strain rate decreases.

At low strain rates (domain III), which depend on the temperature, the processes of athermal strains are dominant: the movement of the dislocations is controlled by the large variations in stress fields induced by the barriers, such as grain boundaries, precipitates and the second phases.

At higher speeds (domain I), the strain mechanisms are thermally activated: the movement of dislocations is increasingly influenced by the small variations in stresses induced by barriers, such as the forest of dislocations and the groups of atoms of the solute in CFC materials, or by the potential of the periodic network (Peierls stress) in CC materials.

In domain II of high deformation speeds, deformation occurs with viscous friction (the dragging of phonons and electrons).

2.5. Mechanisms due to obstacles: hardening mechanisms

For a material to plasticize at a certain strain rate, it is necessary for the force τb (per unit length of the dislocations) to be greater than the resistance, which opposes the movement of the dislocations.

It is necessary to overcome the intrinsic resistances of the bonds between atoms, such as covalent bonds which generate high levels of resistance (in materials such as carbides, oxides, nitrides, etc.). For "ordinary" metals, their stiffness is increased by making them "impure".

The impurities in solution in the metals "counteract" the movement of the dislocations in the sliding planes, and thus increase the resistance. The space between the impurities is related to their concentrations C, and the smaller the spacing between them in the sliding planes, the greater the resistance increases (this is the case for zinc in copper for brass, etc.).

The impurity can precipitate after a resolubilization treatment, once tempered and returned after a controlled time and temperature, in order to obtain an optimum precipitate, giving great resistance to the alloy (this is the case of aluminum alloys with a percentage of copper forming compounds Al_2Cu, steels hardened by carbides, etc.).

Another type of hardening is the work hardening obtained by the increase in dislocations during plastic strain, a type of hardening caused by the interactions between the dislocations, which impede each other's movements and accumulate in the crystals.

Alloy		R_p approx. MP_a	Mechanisms
Aluminium alloys			
AlCuMg with work-hardened treatment	2024 T3	345	Work hardening, precipitation
Annealed AlZnMg	7075-0	100	Solid solution
Treated AlZnMg	7075-T6	500	Precipitation
Copper alloys			
Annealed electrolytic Cu (0.04%O)	–	70	Work hardening
Work-hardened Cu OFHC	–	275	Work hardening
CuZn annealed brass α	–	240	Solid solution
Treated CuBe	–	965	Precipitation
Nickel alloys			
Work-hardened NiCrFe (Inconel)	–	1,035	Work hardening, solid solution
Treated NiMoFe (Hastelloy)	–	275	Solid solution
Treated NiCoCrMoTiAl (Udimet 700)	–	875 (at 650°)	Precipitation
Steels			
Mild steels (0.01%C) annealed	–	170	Solid solution, grain boundaries
Annealed steel (0.2%C)	–	500	Id.
Treated FeNiMoMnCrC	–	1,500	Solid solution, grain boundaries, substructure
Treated maraging	–	2,000	Precipitation, grain boundaries, substructure

Table 2.2. *Hardening mechanisms (François et al. 1991/1992)*

2.5.1. *Obstacles to movement due to the dislocations themselves*

Network forces, recombination, deflected sliding, and jog forming are some of the different obstacles to the movement of dislocations that may result from their own configurations or from their mutual interactions (intrinsic mechanisms).

These athermal processes represent a necessary split τ, which depends little on the speed and temperature. "Thermally activated" processes will result in a split τ, which decreases relatively rapidly when the temperature increases and when the strain rate decreases.

The expression of τ commonly shared among "athermal" processes:

$$\tau \simeq \alpha\, G\, b\, \varrho^{1/2} \simeq \alpha\, G\, b/\bar{\ell}$$

with $\varrho \simeq \bar{\ell}^{-2}$, $\bar{\ell}$ the average distance between the dislocations, or the length of the active segments, resulting from the relationships relating to the stress field of the dislocations and to the line energy (or tension).

To the extent that the density ϱ of dislocations is not modified, the variation of τ on the basis of the temperature is limited to that of the product Gb. On the other hand, the evolution of ϱ with the strain will be the determining factor of the consolidation.

The relationship satisfactorily accounts for the "non-activated part" (at the end of the effects of the near grain boundaries), with α in the order of 0.2 to 0.5: $\tau \simeq 10^{-4}$ to 10^{-3} G.

In the case of the crossing of dislocations, we see the forming of notches. The dislocation, which slips along a plane pierced by another dislocation (this latter dislocation is called a tree), forms a jog of a length equal to b_2 (the screw portion of the Burgers vector of the tree). The displacement of such a jog along the line can occur due to the arrival or by the departure of vacancies. It produces a rise in dislocation.

The energy of formation of a jog is of the order of $0.1\ \mu b^3$.

2.5.2. Interactions with the forest

According to the sign of the dot product $\vec{b_1} \cdot \vec{b_2}$, Burgers vectors of the slip dislocation and the tree, the junction that forms from the crossing is given as:

– attractive $\vec{b_1} \cdot \vec{b_2} < 0$;

– repulsive $\vec{b_1} \cdot \vec{b_2}, > 0$.

At the intersection, the two dislocations can combine to form a dislocation of vectors $\vec{b} = \vec{b_1} + \vec{b_2}$ and energy $\mu/_2\, (\vec{b_1} + \vec{b_2})^2$.

If $\vec{b_1}$ $\vec{b_2}$ is negative, this combination decreases the energy, and it is therefore stable.

As a result, this attractive junction formed in this way can serve as an anchor point for Frank–Read sources.

The attractive junctions are broken for an estimated stress of $\frac{\mu b}{4}\sqrt{\varrho_{disl}}$ for CFC.

Conversely, the crossing of the trees exerting a repelling force only requires the creation of a jog, which represents a much lower force; for a jog energy of $0.1\ \mu b^3$, the necessary stress is of the order of:

$$\tau \simeq \frac{\mu b}{30}\sqrt{\varrho_{disloc}}$$

We therefore have a stress that we must exert to drag the dislocation in the middle of the forest; we have a critical forest τ equal to:

$$\tau = \alpha\mu b\sqrt{\varrho_{disloc}}$$

with:

$$\propto\ \simeq 1/3 \text{ to } 1/4$$

This stress increases as the dislocations multiply during the plastic deformation; this explains the strain hardening through the corresponding work hardening.

2.5.3. *(Simplified) analysis of the consolidation due to the forest dislocation*

Work hardening, considering the influence of the average free path ΔL of the mobile dislocations ϱ_m in the forest ϱ_F on $\sigma = f(\varepsilon)$, is given as:

$$\varepsilon = \varrho_m\ b\ \Delta L \qquad\qquad\qquad [2.7]$$

and:

$$\dot{\varepsilon} = \varrho_m\ b\ v \qquad\qquad\qquad [2.8]$$

v: speed of mobile dislocations of Burgers vector B.

We obtain:

$$d\varepsilon = d\varrho_m b\ \Delta L \qquad\qquad\qquad [2.9]$$

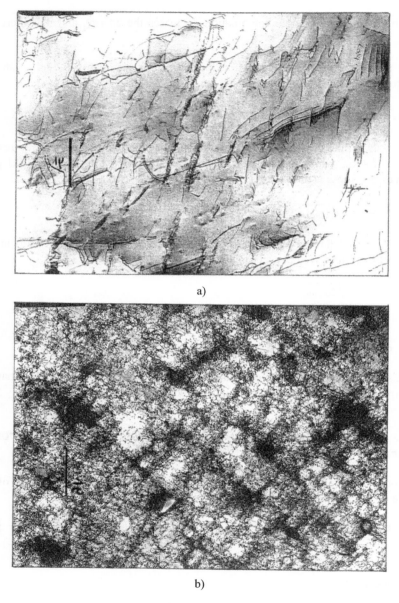

a)

b)

Figure 2.8. *Multiplication of dislocations during strain and work hardening (CEA-SRMA photos) (the line represents one micrometer). (a) Dislocations in an annealed Hastelloy alloy. (b) After a work hardening of 15%, the dislocations begin to form clusters*

We have:

$$\varrho_m = k \, \varrho_F \qquad [2.10]$$

and:

$$\sigma = \frac{\mu b}{\beta \ell} \qquad [2.11]$$

with:

$$\beta \ell = \varrho_F^{-1/2} = \left(\frac{\varrho_m}{k}\right)^{-1/2} \qquad [2.12]$$

[2.11] and [2.12]:

$$\sigma = \frac{\mu b}{\beta \ell} = \frac{\mu b}{\beta}\left(\frac{\varrho_m}{k}\right)^{+1/2}$$

Which gives:

$$\varrho_m = k\left(\frac{\beta \sigma}{\mu b}\right)^2 \qquad [2.13]$$

and [2.9]:

$$\frac{d\varrho_m}{d\varepsilon} = \frac{1}{b \Delta L} = \frac{d}{d\varepsilon}\left[k\left(\frac{\beta \sigma}{\mu b}\right)^2\right]$$

we obtain:

$$\frac{d}{d\varepsilon}\left(\frac{\sigma}{\mu}\right)^2 = \frac{b}{\beta^2 \, k \, \Delta L} \qquad [2.14]$$

We will now consider the important parameter ΔL; the average free path of mobile dislocations is given as:

– if ΔL is constant, then $\left(\frac{\sigma}{\mu}\right)^2 \propto \varepsilon$. The hardening is parabolic;

– for $\Delta L \propto \frac{1}{\varepsilon}, \frac{\sigma}{\mu} \propto \varepsilon$, the hardening is linear.

For a material whose behavior is $\sigma = A\varepsilon^n$, the consolidation is equal to:

$$n = \frac{d \, Ln \, \sigma(\varepsilon)}{d Ln \varepsilon} \qquad [2.15]$$

Metal	n
Extra-mild steel	0.15–0.25
17% Cr ferritic steel	0.16–0.20
Austenitic steel 18-10	0.40–0.50
Aluminum	0.07–0.27
AG 1 to 5	0.23–0.30
A S G	0.23
AU 4 G	0.15
Brass 67 Cu-33 Zn	0.55
Brass 63 Cu-37 Zn	0.45
Copper	0.30–0.47
Zinc	0.1
Nickel	0.6

Table 2.3. *Values of N for different materials (Pomey, Grumbach)*

2.5.4. *Blockages of dislocations by pileups*

Let us assume that a Frank–Read source produces many dislocations located on a sliding plane: the dislocations may encounter obstacles, which stop their movements, such as a grain or sub-grain boundary. The dislocations located behind the dislocation immobilized at the grain boundary will pile up on it and may remain blocked if they are unable to move within another plane (a thermal aid may cause a climbing movement).

The dislocations are subjected to a shear stress τ due to mutual repulsive forces and the action of the obstacle. It can be found at the positioning equilibrium that the resultant force acting on each dislocation is zero, and for the n dislocations:

$$\frac{\mu b^2}{2\pi (1-v)} \sum_{\substack{j=0 \\ j \neq i}}^{n} \frac{1}{x_i - x_j} - \tau b = 0 \tag{2.16}$$

(if it is a screw dislocation, it is necessary to remove the factor $1 - v$).

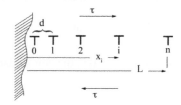

Figure 2.9. *Piling-up of dislocations*

2.5.5. *Piling-up of dislocations*

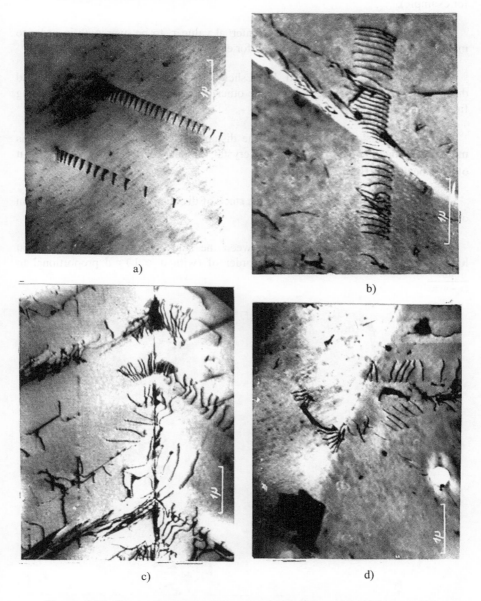

a)

b)

c)

d)

Figure 2.10. *Cupro-aluminum. Piling-up of dislocations (a) UA8 to ε = 4%.*
(b) and (c) UA6 Mn2. (d) UA6 (Mason 1968; Barnouin 1972)

The x_i, ... are the distances of the dislocations to the barrier (the boundary plane, for example).

τ is the stress after subtracting the internal value τ_i to be overcome in order to move the dislocations (the Peierls force, for example).

Equation [2.16] was solved by Eshelby, Franck and Nabarro (the head dislocation undergoes a stress due to the other dislocations and to the stress arising from the applied shear).

For the nth dislocation located at the distance L from the barrier, and if L is much lower than the dimension D of the crystal, the energy of a stack is close to that of a dislocation of Burgers vector $n\vec{b}$.

As a first approximation, we equate a stack of dislocations to a super-dislocation of Burgers vector $n\vec{b}$.

The evaluation of the distance between the stack head dislocation and that located immediately behind is of the order of (with d ≪ L) : d proportional to $\dfrac{\mu b}{(1-v)\, n\tau}$.

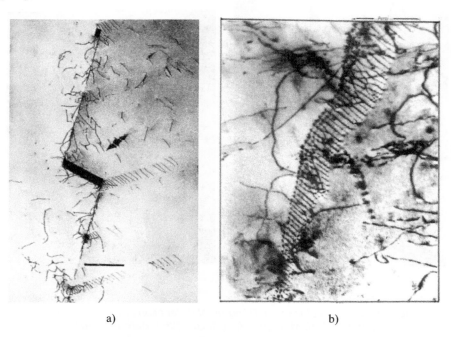

a)   b)

Figure 2.11. *Piling-up of dislocations*

COMMENT ON FIGURE 2.11.– *(a) Stacks of dislocations in stainless steel of 18-10. They are visible on the right part of the shot. The sliding planes are inclined relative to the thin blade and the dislocations pass through it (the line represents a micrometer) (photo: CEA SRMA). (b) Direct electron micrography of a sample of 15% Al-MG alloy, cast and thinned electrolytically. This range represents two subgrains which occupy the right half and the left half of the figure, respectively, and which are separated by a wall in which the individual dislocations are apparent. This wall is arranged in depth in the sample (photo: Saulnier, Trillat).*

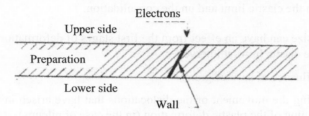

Figure 2.12. *Wall of the two subgrains where the dislocations appear*

2.5.6. *Influence of grain size*

The experimental curves show that the grain size acts on the yield strength, and possibly on the consolidation (Figure 2.13).

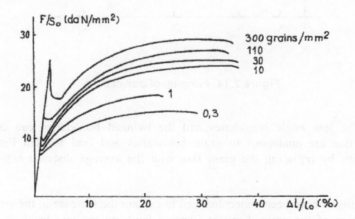

Figure 2.13. *Tensile curves of iron with 0.02% carbon at ambient temperature as a function of the average grain diameter (Bouchet et al.)*

The general ratio that gives the evolution of the stress $\sigma(\epsilon)$ as a function of the grain diameter is the "Petch's law":

$$\sigma(\epsilon) = \sigma_i(\epsilon) + k(\epsilon).d^{-1/2}$$

$- k(\epsilon)$ is not affected very strongly by the imposed deformation conditions (speed and temperature);

$- \sigma_i(\epsilon)$ changes in accordance with the influence of the speed $\dot{\epsilon}$ or the temperature on the elastic limit and on the consolidation.

The grain size can have an effect from the first stages of deformation and modify the value of the elastic limit, it intervenes:

– by the influence of surface grains;

– by blocking the movement of the dislocations that have arisen in the material, from the beginning of the plastic deformation (in the case of pileups).

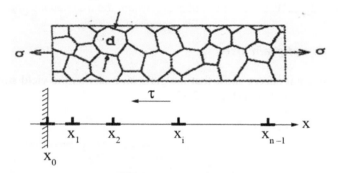

Figure 2.14. *Piling-up of dislocations*

NOTE.– The low angle boundaries and the twinned boundaries can constitute obstacles that are analogous to grain boundaries and lead to Hall–Petch type relationships by replacing the grain size with the average distance between the obstacles.

Mechanisms have been conceptualized to explain the increase in the stress $\sigma(\varepsilon)$ as a function of the grain diameter (Petch's law): on the one hand, due to the disorientation between crystals, the sliding planes of the adjacent crystals do not coincide; on the other hand, the boundaries are regions where the arrangement of the atoms is disturbed and requires readjustments (dislocations of boundaries), and where the concentrations of impurities or elements of alloys can be much higher (boundaries are often preferred sites for the formation of precipitates). Figure 2.15

indicates the various mechanisms with the potential to interact, and Table 2.4 gives the corresponding K values.

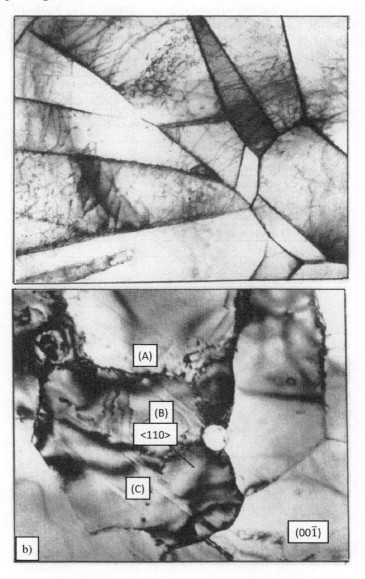

Figure 2.15. *Subgrains and dislocation walls*

COMMENT ON FIGURE 2.15.– *Obstacles: sub-grains separated by dislocation walls, direct electron micrographs of two electrolytically thinned samples, the observation ranges are monocrystalline and divided into sub-grains. In (a) hardened beryllium (x 26,000). In (b) 1.2% work-hardened Al-Mn alloy (x 65,000) (plane of figure (00$\bar{1}$), we observe the displacements of dislocations in the form of ribbons (A), (B), (C), etc., parallel to the directions $\langle 110 \rangle$ (Saulnier, Trillat).*

	(daN/mm^2)	K (daN/mm$^{3/2}$)
Iron	3.6–4.5	0.66–2.3
Niobium	6.9	0.13
Vanadium	3.09	0.72
Molybdenum	10.8	5.7
Copper	2.7	0.4
Aluminum	1-1.5	0.2
Nickel	3.6	0.1
Magnesium	0.7	0.9
Titanium	13–47	0.98–0.48
Brass	2.8-4.5	1-1.3
Fe-Si (6 %)	30	0.96
Fe-Co (ordered)	5	0.96
Fe-Co (not ordered)	37.8	0.4

Table 2.4. *Values of σ_i and k of different metals at the elastic limit (Hall, Morrison, Leslie)*

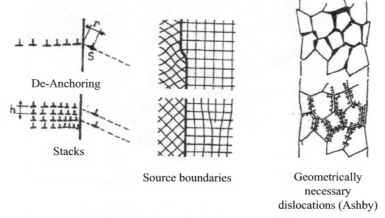

De-Anchoring

Stacks

Source boundaries

Geometrically necessary dislocations (Ashby)

Figure 2.16. *Types of strengthening mechanisms for the boundaries*

De-anchoring of blocked sources – stacking effects	$\sigma_c.r^{1/2}$ or $(A\sigma_c)^{\frac{1}{2}}$	$r \simeq$ some b $A \cong \frac{\beta Gb\,(2-\nu)}{1+\upsilon}$ or $\delta.G\frac{b}{h}$ $\beta \simeq \delta \simeq 0.1\text{–}0.6$
Source boundaries	$\alpha Gb\rho_s^{1/2}$	$\rho_s =$ density of sources at the boundaries
Accumulation of dislocations in the crystals Geometrically necessary dislocations	$\alpha G(b\epsilon)^{1/2}$	$\alpha \simeq 0.1\text{–}0.4$

Table 2.5. Effect of hardening owing to the boundaries[2]

σ_c represents the shear necessary to unblock a source of dislocations at or near the boundary, subjected to actions by piles of dislocations in the neighboring grain.

NOTE.– The increase in the density of dislocations during deformation can create "cell arrangements" whose diameter decreases, and the shape changes with deformation.

Petch's law is confirmed if we observe the stress $\sigma(\epsilon)$ in contrast with the average diameter of the cells or their minimum dimension at power $-1/2$ (this is the case for mono and polycrystals). In particular, this fact explains the decrease in the coefficient $K(\epsilon)$ joint consolidation when ϵ increases, with the action of the cell walls replacing that of the boundaries.

2.5.7. The case of nanomaterials (very small grain size): mechanical strength and hardness (T less than 0.3 Tf)

The classical Hall–Petch ratio expresses the effect of the grain size on the plastic flow rate of polycrystalline materials, whether or not they are work-hardened. Explained for the apparent yield strength σ or for the hardness H, it becomes:

$$\sigma = \sigma_0 + k\,d^{-0.5},\ H = H_0 + K_H.d^{-0.5}$$

---

2 Armstrong, R.W. (1970). The influence of polycrystal grain size on several mechanical properties of materials. *Metall. Trans.*, 1, 1169; Li, J.C.M. and Chou, Y.T. (1970). The role of dislocations in the flow stress grain size relationships. *Met. Trans.*, I, 1145; Ashby, M.F. (1970). The deformation of plastically non-homogeneous materials. *Phil. Mag.*, 21, 399; Li, J.C.M. (1963). Petch relation and grain boundary sources. *Trans. Aime*, 227, 239.

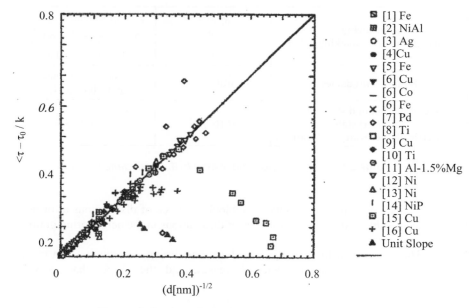

Figure 2.17. *Hall–Petch ratio (Masumura et al. 1998)*

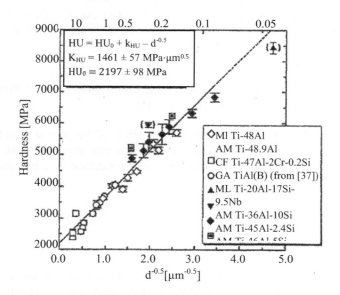

Figure 2.18. *Dureté (Bohn et al. 2001)*

EXAMPLE 2.1.–

Evolution of the elastic limit and stress in relation to grain diameter.

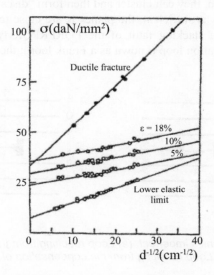

Figure 2.19. *Soft iron (Armstrong 1970)*

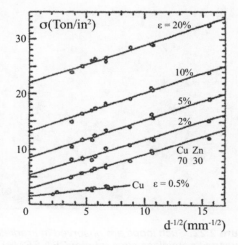

Figure 2.20. *Copper and Cu-Zn alloy (70-30)*
1 Ton/in^2 ≈ 1.57 daN/mm^2 (Meaking and Petting 1974)

2.5.8. *Influence of vacancies*

A crystal always contains vacancies. If these are in supersaturation as a result of tempering or irradiation, they can cluster and then form "discs" along a dense plane. If the size of the disks is sufficient, the planes R are close to the planes P (Figure 2.21), which creates a stacking fault of the PQRPRPQR type, surrounded by a so-called sessile dislocation loop (known as a Frank loop); the stacking faults create barriers for the slip.

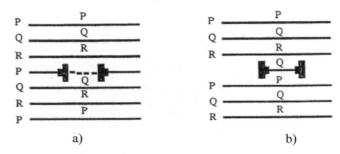

a)   b)

Figure 2.21. *Frank's imperfect. (a) Loop resulting from the condensation of gaps. (b) Loop resulting from the condensation of interstitials*

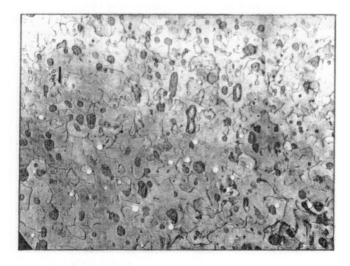

Figure 2.22. *Such loops are observed in irradiated austenitic stainless steel (photo: CEA SRMA)*

2.5.9. *Vacancies and dislocations*

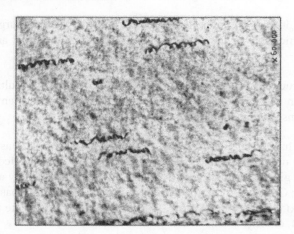

Figure 2.23. *Electronic micrograph (60,000×) of a 4% Al-Cu alloy
tempered at 525°C and thinned electrolytically (with a thin blade), dislocations
in helices resulting from the interaction of linear dislocations and gaps*

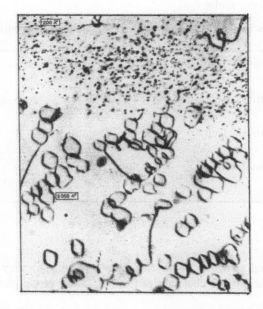

Figure 2.24. *Electronic micrograph (65,000×) of a 1.2% Al-Si alloy
tempered at 570°C. The vacancies accumulated in loops of dislocations
between 200 and 2000 Å (photo: Saulnier and Trillat)*

2.5.10. *Stacking and twinning faults*

Stacking faults play an important role in the study of the metallurgical properties of crystals, particularly in the analysis of dissociated dislocations and mechanical faults.

In metal structures, there may be several types of stacking faults, for which a brief description may prove useful. A crystal in a compact phase can be considered as a stack of dense atomic planes.

Each plane has a triangular structure. By projecting the positions of the atomic sites of a plane on a given plane, three possibilities exist, represented by the letters A, B, and C. A given structure (with or without faults) is defined by a succession of planes A, B, or C. We also use the symbol Δ to indicate that the passage from one plane to the next is of the type AB (or BC, or CA), and ∇ to indicate that it is of the type BA (or CB, or AC).

A perfectly cubic crystal with centered faces is characterized by a stack of planes whose arrangement repeats periodically: ... ABC ABC ... The introduction of an "intrinsic" stacking fault produces an error in this sequence; for an "extrinsic" fault, we find two successive faults.

– For a spinel twin, we see a stacking fault for each plane in half of the crystal:

Perfect crystal	A B C A B C Δ Δ Δ Δ Δ
Intrinsic fault	A B C ↓ B C A Δ Δ ∇ Δ Δ
Extrinsic fault	A B C ↓ B ↓ A B Δ Δ ∇ ∇ Δ
Spinel twin	A B C ↓ B ↓ A ↓ C Δ Δ ∇ ∇ ∇ ∇

Table 2.6. *CFC Structure*

– The hexagonal structure (with two atoms per mesh) may or may not be compact, and its defects are characterized by the stacks of the planes shown in the table.

The dislocation line limits the part of the crystal that has slipped relative to the one that has remained in place. The amount of slip that occurs is B. It is possible to consider cases where this quantity is not a vector of the network, but is instead a smaller vector. Thus, the Burger vector dislocation b can be "dissociated" via a reaction $b \rightarrow b_1 + b_2$ in two "partials" which limit a "stacking fault"; here the crystal is imperfect along the surface which separates the two partials.

Perfect crystal		B		A		B		A		B		A	B
			∇		Δ		∇		Δ		∇		Δ
Fault 1 Δ or twin		B		A		B ↓	C		B		C		B
			∇		Δ		Δ		∇		Δ		∇
Fault 2 Δ intrinsic		B		A		B ↓	C ↓	A		C		A
			∇		Δ		Δ		Δ		∇		Δ
Fault 3 Δ extrinsic		B		A		B ↓	C ↓	A ↓	B		A
			∇		Δ		Δ		Δ		Δ		∇

Table 2.7. *Hexagonal structure*

A perfect dislocation L of the cubic system with centered faces can break down, giving two Shockley dislocations $\ell\ell'$ separated by a stacking fault in the plane (111):

$$\frac{a}{2}[110] \rightarrow \frac{a}{6}[121] + \frac{a}{6}[21\overline{1}]$$

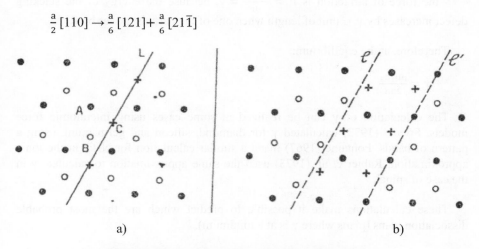

a)   b)

Figure 2.25. *Decomposition into half-dislocations in the cubic system with centered faces. Projection on a plane (111)*

The decomposition amounts to replacing a row L of additional atoms C (Figure 2.25(a)) with a band $\ell\ell'$ of atoms C in the absence of stacking on the substrate of atoms A (Figure 2.25(b)).

The dislocation is associated in two "sliding" partials (or Shockley partials) contained in the slip plane, referred to as such because they allow the dislocation to slip. If the partials have a Burgers vector that is not in the slip plane, they are "sessile" (Frank partials), because they do not allow the dislocation to slip.

If we consider the dissociation from the point of view of energy, the non-dissociated dislocation has an energy of $E \simeq \mu b^2$, and the two partials have an energy of $E' \simeq \mu (b_1^2 + b_2^2)$. Subsequently, a dissociation is likely if $b_1^2 + b_2^2 < b^2$ (Frank's criterion).

However, this criterion does not provide sufficient conditions. It will then be necessary to take into account the energy of the fault γ (energy/cm^2).

Let us consider a dissociation $b \rightarrow b_1 + b_2$, assuming screw dislocations and d the distance between partials:

– the repulsion force of the two partials is $F = \frac{\mu b_1 b_2}{2\pi d} \ (\vec{b_1}\vec{b_2} > 0)$;

– the force of attraction is $F = \frac{\partial(\gamma.r)}{\partial r} = \gamma$, because the energy of the stacking defect increases by $\gamma . r$/ unit of length when one of the partials moves by R.

Therefore, at the equilibrium:

$$\gamma = \frac{\mu b_1 b_2}{2\pi d}$$

The calculation of γ can be realized in some cases using interatomic force models. Friedel (1972) calculated γ for diamond, silicon and germanium, using a pattern of bands. Fontaine (1967) made a similar calculation for NaCl in the ionic approximation. Rabier et al. (1973) used the same approximation to calculate γ in the case of spinel.

These calculations make it possible to predict which are the most probable dissociation plans (plans where γ is at a minimum).

Analogous breakdowns can occur in other crystal systems. Table 2.8 gives the results for a few simple systems: cubic, diamond-type CD; hexagonal and compact HC; centered cubic CC; centered faces and cubic CFC.

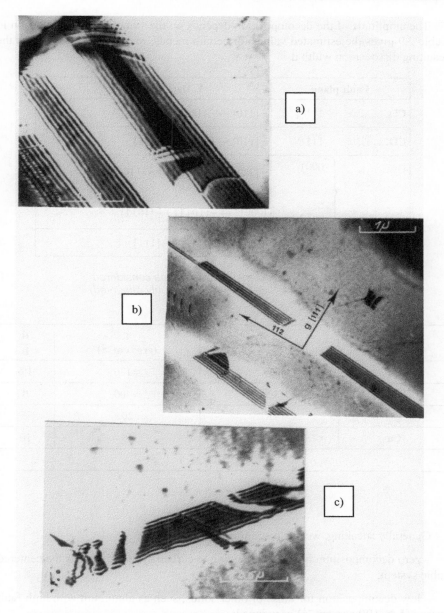

Figure 2.26. *Cupro-aluminum alloy stack faults UAB (a) deformed by 4%, in superposition; (b) in annealed state; (c) deformed by 4% (Barnouin 1972)*

The amplitude of the decomposition depends on the value of the fault tension F. Table 2.9 gives the estimated values for certain metals, in ergs.cm^{-2}, as well as the resulting dissociation width d.

Fault plane		Decomposition
CFC.........	111	$\frac{1}{2}[110] \rightarrow \frac{1}{6}[121] + \frac{1}{6}[21\bar{1}]$
CD..........	111	$\frac{1}{2}[110] \rightarrow \frac{1}{6}[121] + \frac{1}{6}[21\bar{1}]$
HC..........	0001	$\frac{1}{2}[2\bar{1}\bar{1}0] \rightarrow \frac{1}{2}[10\bar{1}0] + \frac{1}{2}[1\bar{1}00]$
	$10\bar{1}2$	$[0001] \rightarrow \frac{1}{2}[\bar{1}011] + \frac{1}{2}[10\bar{1}1]$
CC..........	112	$\frac{1}{2}[111] \leftarrow \frac{1}{3}[112] + \frac{1}{6}[11\bar{1}]$

Table 2.8. *Fault plans (the fault plans considered are those for which twins are commonly observed)*

Metal	System	f (ergs.cm^{-2})	$\dfrac{d}{b}$
Al	CFC	$\simeq 170$	1.6
Fe Si	CC	$\simeq 400$	0
Co	HC	20	35
Cu	CFC	40	10
erg: 10^{-7} J			

Table 2.9. *Fault tensions*

Generally speaking, we can distinguish three cases:

– zero decomposition in solids belonging to certain systems, such as the centered cubic system;

– low decomposition for those of other solids, such as aluminum with high fault energy (f > 100 erg. cm^{-2}) or twinned;

– high decomposition for the remaining solids (suitable crystal systems, f < 100 erg. cm^{-2}). The last of these categories is likely to contain solids that easily form annealing stains; that is, in addition to copper and cobalt, iron γ, silver and the solid solutions AlSi, CuZn, CuSi, cubic solids of the diamond type, etc.

2.5.11. *Lomer–Cottrell barriers*

Cottrell has shown that if there are edge dislocations located in differently oriented planes that intersect in the crystal, two of these dislocations can be combined, giving imperfect Shockley dislocations. These Shockley dislocations, involving stacking defects belonging to two plane systems, cannot leave the intersection of these planes and therefore cannot move.

In this way, we can explain the formation of "barriers" preventing the displacement of dislocations from Frank–Read sources. These dislocations accumulate as the deformation progresses upstream of the barriers, creating zones of energy concentration. The displacement of the following dislocations becomes more and more difficult and the critical shear stress of the crystal increases; this is "consolidation".

In any case, the critical shear stress increases as the density of the dislocations decreases, and it is assumed that the critical shear stress will depend on the density of the dislocations according to a relationship of the following form:

$$\tau = \tau_0 + A\,G\,b\,\sqrt{\varrho}$$

where τ_0 is the critical shear stress before consolidation, A is a constant in the order of 0.3–0.6, G is the sliding modulus, b is the Burgers vector and ϱ is the density of dislocations.

2.5.11.1. *The case of CFC monocrystals*

To explain the hardening of cubic monocrystals with centered faces, it is necessary for the crystal to have relatively fixed "barriers", which are capable of retaining part of the dislocations that occur during the deformation. The Frank sessile dislocations may act as such barriers, but it is difficult for them to form.

Cottrell (1952), developing an idea from Lomer (1951), introduced another type of barrier that is much easier to form in the cubic system with centered faces.

Consider two dislocation lines of Burger vectors $\frac{a}{2}$ [011] and $\frac{a}{2}$ [10$\bar{1}$], located in two different slip planes (11$\bar{1}$) and (111), and both parallel to the intersection [1$\bar{1}$0] of these planes. If these dislocations are broken down into half-locations,

$\frac{a}{6}$ [$\bar{1}$21] and $\frac{a}{6}$ [112], $\frac{a}{6}$ [2$\bar{1}\bar{1}$] and $\frac{a}{6}$ [11$\bar{2}$], these attract each other from one dislocation to another (Figure 2.27(a)), and the dislocations therefore approach the intersection [1$\bar{1}$0] until the half-dislocations on the left side combine, following the reaction:

$$\frac{a}{6}\,[\bar{1}21] + \frac{a}{6}\,[2\bar{1}\bar{1}] \rightarrow \frac{a}{6}\,[110]$$

The imperfect dislocation formed in this way is stable: it sufficiently repels the remaining half-dislocations $\frac{a}{6}$ [112] and $\frac{a}{6}$ [11$\bar{2}$] to counterbalance their mutual attraction (Figure 2.27(b)). It belongs to the stacking faults of both the planes (111) and (11$\bar{1}$), and therefore cannot leave their intersection [1$\bar{1}$0]. This grouping is therefore immobile. Many barriers of this type have been observed by Jacquet in brass α, at the intersection of two slip lines (1954).

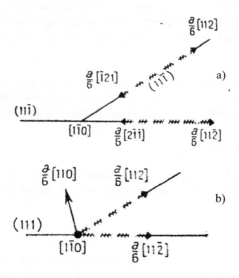

Figure 2.27. *Cottrell dislocations. (a) Initial state. (b) Barrier*

The case studied by Cottrell is the only one where dislocations attract at great distances and form a stable barrier (Friedel 1955).

The Cottrell barrier is also the most stable of the possible barriers.

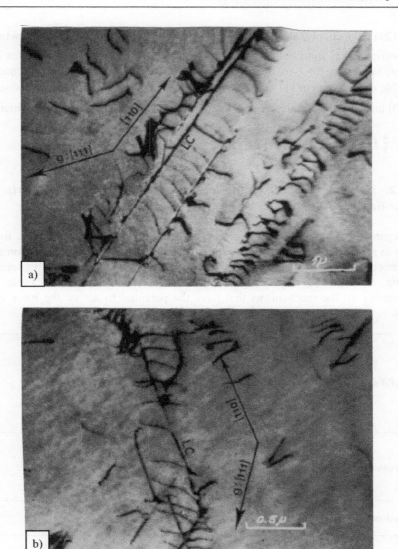

Figure 2.28. *Electron micrography of UA6 Mn2 in (a) and (b): Lomer–Cottrell barriers (Barnouin 1972)*

2.5.11.2. *CC single crystal cases*

A barrier of the same type can occur in the centered cubic system: two dislocations with Burgers vectors $\frac{a}{2}[11\bar{1}]$ and $\frac{a}{2}[1\bar{1}1]$, located in the planes (112) and (121) respectively, attract each other to form a dislocation $\frac{a}{2}$ [200] following the intersection [$3\bar{1}\bar{1}$] from both planes. The resulting dislocation then dissociates into a barrier, consisting of two bands of stacking faults, bordered in both planes by movable dislocations of [$11\bar{1}$] and $\frac{a}{6}$ [$1\bar{1}1$] and separated by an immobile dislocation $\frac{a}{3}$ [200] located at the intersection of the planes. The series of reactions is written as:

$$\frac{a}{2} [11\bar{1}] + \frac{a}{2} [1\bar{1}1] \rightarrow \frac{a}{2} [200] \rightarrow \frac{a}{3} [200] + \frac{a}{6} [11\bar{1}] + \frac{a}{6} [1\bar{1}1]$$

2.5.12. Influence of obstacles associated with alloy elements and impurities

Regardless of the effect of the addition elements on the dissociation, foreign atoms in solid solutions (insertion or substitution), clusters or Guinier–Preston zones and precipitates constitute obstacles that are more or less effective in terms of moving dislocations, depending on their size and distribution. Table 2.10 summarizes all the mechanisms that have the potential to act on the basis of the alloys.

Solid solutions	Size effect
	Pinning effect
	Elastic modulus effect
	Asymmetry effect
	Segregation effect
	Order effect
Areas of coherent precipitates	Size effect
	Modulus effect
	Binding effect
	Order effect
Non-coherent precipitates	Anchoring effect
	Dislocation multiplication effect

Table 2.10. Effects of obstacles to the movement of dislocations due to inclusions and elements of alloys

The presence of elements increases the resistance of polycrystalline materials; the dislocations must be able to cross them by shearing, bypassing (as is the case with precipitates):

– τ_0 increases with the content of impurities, at least at low concentrations;

– at room temperature, τ_0 varies following a linear pattern as a function of the concentration, but the rate of variation of τ_0/dc depends on the addition element (solute);

– in logarithmic coordinates, $d\tau_0/dc$ varies linearly depending on the difference ΔD of the atomic diameters of the solvent (copper) and the solutes.

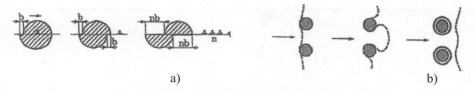

a)   b)

Figure 2.29. *Interactions of dislocations with impurities (a) by shearing and (b) bypass with the formation of curls*

2.5.13. *Anchoring dislocations*

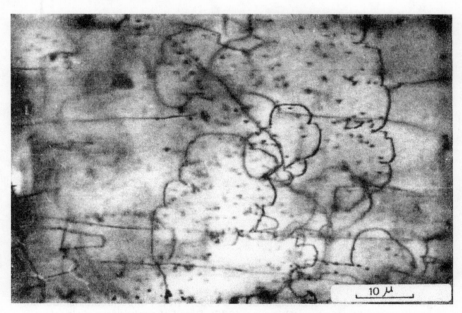

Figure 2.30. *Precipitate in a neoblast and anchoring dislocations (Gueguen 1976)*

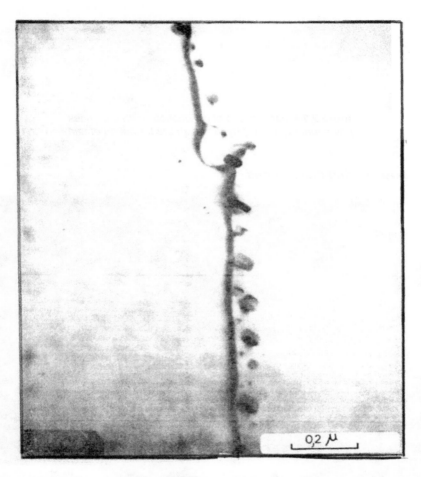

Figure 2.31. *Presence of precipitates in a nodule (rock) of peridotite, kimberlite, prophyroclastic structure (Gueguen 1976). Detail of anchoring a dislocation on the precipitates (electron microscopy of 125 kV × 80,000)*

2.5.14. *Formation of loops around particles*

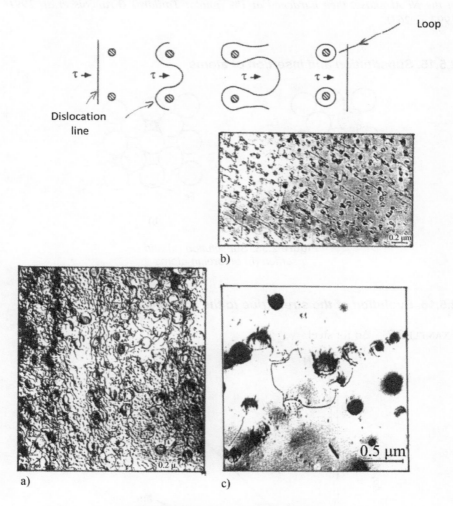

Figure 2.32. *Orowan's mechanism*

COMMENT ON FIGURE 2.32.– *(a) The image shows dislocation loops that are left around the particles γ in a nickel-based alloy, Waspaloy (18.7 Cr-14 C_0-3 Ti-1.45 Al-3.9 Mo the rest (weight %), cyclically deformed ($\Delta\varepsilon_p/2 = 0.2\%$) at 650°C. (photo: Clavel). (b and c) Dislocation loops left around the particles in ferritic* *stainless steel (Fe Cr Ni Al) aged (650°C – 6 h) and hardened through precipitation of the Ni Al phase, then hardened at 1% (photo: Taillard) (François et al. 1991/ 1992, p. 263).*

2.5.15. Substitution and insertion of atoms

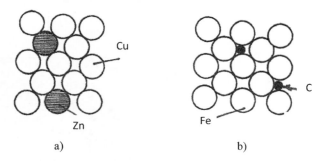

a)   b)

Figure 2.33. *Substitution (a) and insertion (b) of foreign atoms*

2.5.16. Evolution of the stress due to the added elements

EXAMPLE 2.2.– Δσ for steel.

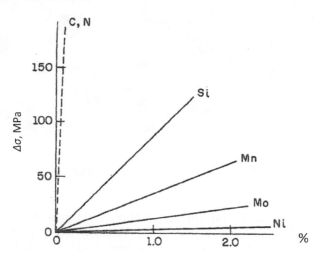

Figure 2.34. *Increase Δσ in steel according to the percentage of additional elements in substitution (solid line) and in insertion (dotted lines) (according to Pickering and Gladman, ISI Special Report 81, Iron and Steel Inst., London, 1963)*

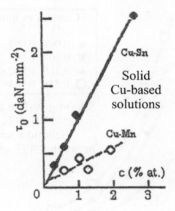

Figure 2.35. *Linear evolution of τ_0 on the basis of the concentration of additional elements: the case of the Cu with Sn and Mn*

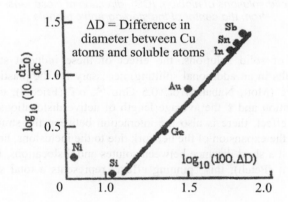

Figure 2.36. *Linear variation of $d\tau_0/DC$ as a function of ΔD (in log)*

NOTE.– According to de Fouquet (1976), the effect of the size on its general shape is due to the difference in volume between the "inclusion" and the domain of the network it occupies, thus causing the stress induced by this difference. r is the radius of the inclusion and η is the size factor $\eta \simeq |\Delta|/r$ for an inclusion located in a radius site $(r - \Delta)$ from the initial network.

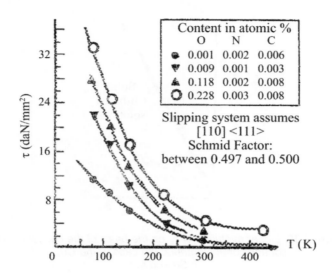

Figure 2.37. *Conventional critical split resolved (ε = 0.1 %) of solid solutions of niobium (CC), direction of close tension from the center of the stereographic triangle*

In the case of solid solutions, the effect of these induced stresses on the dislocations results in an additional splitting, necessary for the plastic strain of the form: $\Delta\tau \simeq G\eta^2 c$ (Mott, Nabarro) or: $0.5\ G\eta c^{2/3}$. b/ℓ (Friedel); c indicates the atomic concentration and ℓ the average length of active dislocations. However, in addition to this effect, there is also the interaction between the strain field of the dislocations and the expansion of the network due to the inclusions, hence an energy of "interaction" at a short distance between solutes and dislocations, in the form of: $W_m = 0.5\ Gb^3.\eta$ (Petch); this "pinning effect" represents a total split $\Delta\tau$ in the form of:

$$\Delta\tau = \Delta\tau_0 + \Delta\tau_c\ (1 - T/T_c)\ \text{with}\ \Delta\tau_0 \simeq 0.25\ G\ (\eta c)^{4/3}$$

and $\Delta\tau_c \simeq 0.25\ G\eta c$ (Friedel, Mott). As for the intrinsic activated processes, T_c is an increasing function of the deformation rate.

The local variations in elastic stresses due to the atoms in solution can also contribute to the increase in the splitting $\Delta\tau$. Several different expressions have been proposed to take into account the combined effect of size and modulus. All these expressions represent "hardening" in the form $\Delta\tau \simeq G\eta.c$ with η between 1/2 and 1.

Fleisher has noted that it is in fact possible to distinguish between two types of solid solution hardenings: a weak hardening, such that $d\tau/DC \simeq G/10$ to $G/20$ corresponding to symmetrical inclusions (substitutional, interstitial in CFCs), and a strong hardening, such that $d\tau/DC \simeq G/2$ to $G/3$ corresponding to asymmetric inclusions (interstitial in the CC). In this second case, the Fleisher model provides for a hardening in the form:

$$\Delta\tau \simeq \Delta\tau_c\left\{1 - (T/T_c)^{1/2}\right\}^2 \text{ with } \Delta\tau_c \simeq \propto G\,(\Delta\varepsilon)c^{1/2}$$

$\Delta\varepsilon$ characterizes asymmetry and $\propto \simeq 1/3$ to $1/4$.

Schoeck considered the possible shuffling of the interstitials in a CC network when a dislocation occurs nearby, leading to a hardening $d(\Delta\tau)/dc = 65$ daN/mm^2 for the carbon in the iron.

2.5.16.1. *Simplified analysis of dislocation–inclusion interaction*

The case of an edge dislocation with its own stress field (hydrostatic pressure σ_p) and the expansion of the network due to inclusion (change in volume ΔV).

EXAMPLE 2.3.–

The edge dislocation has a stress field that has been established earlier (Chapter 1, section 1.3.2.3). For example, using cylindrical coordinates, we obtain:

$$\sigma_{rr} = \sigma_{\theta\theta} = \frac{\mu b}{2\pi\,(1-\nu)}\frac{\sin\theta}{r}, \text{ etc.}$$

Given that the value of σ_p is:

$$1/3\,(\sigma_{rr} + \sigma_{\theta\theta} + \sigma_{zz})$$

we obtain:

$$\sigma_p = \frac{1+\nu}{1-\nu}\frac{\mu b}{3\pi}\frac{\sin\theta}{r}$$

In linear coordinates (x, y, z), with $r = (x^2 + y^2)^{1/2}$ and:

$$\text{Sin}\,\theta = y/(x^2 + y^2)^{1/2}$$

We obtain $\sigma_p\,(x, y, z)$.

If we take an expanded volume V of ΔV by inclusion, and considering a deformation of ε (low) for the radius r_0 of the cavity, the change in volume is:

$$\Delta V = \frac{4}{3} \pi (r^3 - r_0^3) = \frac{4}{3} \pi r_0^3 (1 + \varepsilon)^3 - \frac{4}{3} \pi r_0^3$$

$$\Delta V \simeq \frac{4}{3} \pi r_0^3 \, 3\varepsilon = 4 \pi r_0^3 \, \varepsilon$$

The interaction energy W is close to:

$$W = \sigma_p \, \Delta V$$

$$W = \underbrace{\frac{4}{3} \frac{1+v}{1-v} \mu \, b \, \varepsilon \, r_0^3}_{K} \cdot \frac{\sin \theta}{r}$$

giving a force F on the dislocation:

$$F = \partial W / \partial r = K \sin \theta / r^2$$

For inclusions with a spacing of d with the pinning effect of the dislocations, for dislocations of length d, we obtain:

$$\Delta \tau = F/bd$$

In the case of a concentration c of the inclusions per unit volume and if ϱ is the density of dislocations (total length of dislocations per unit volume):

$$d = \varrho/c$$

We obtain:

$$\Delta \tau = K \sin \theta \, c / r^2 \, b\varrho$$

If $r \simeq b$ and, for $\sin \theta = 1$, $\Delta \tau$ is in the order of:

$$\Delta \tau \simeq K c / b^3 \, \varrho$$

The value represents the maximum of the interaction.

NOTE.– The pinning and modulus effect models primarily apply to substitutional solid solutions and CFC interstitial solid solutions. On the other hand, the effects of tetragonality and induced rearrangement explain the strong hardening of the solid interstitial CC solutions.

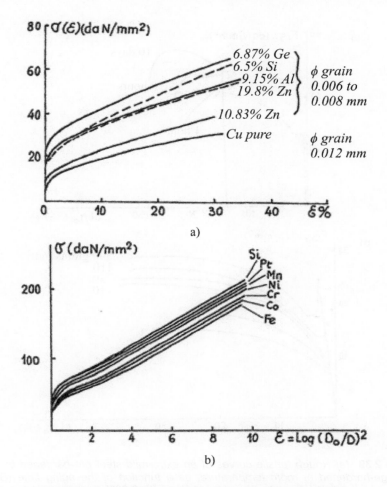

Figure 2.38. *(a) Rational tensile curves of copper and some of its alloys (McLean). (b) Consolidation curves of non-alloyed and alloyed iron (c = 3%) as a function of the stretching in true strain (Leslie)*

NOTE.– In addition to these "average" effects, the "segregation" of the addition elements or impurities may occur along the dislocations, which is responsible for the change in the mechanical characteristics of the extra-mild steels, particularly during age hardening, and the existence of a more or less strong "upper" elastic limit. The critical split level τ_c needed to "free" the dislocations in this case is in the order of: $\tau_c \simeq G/50$.

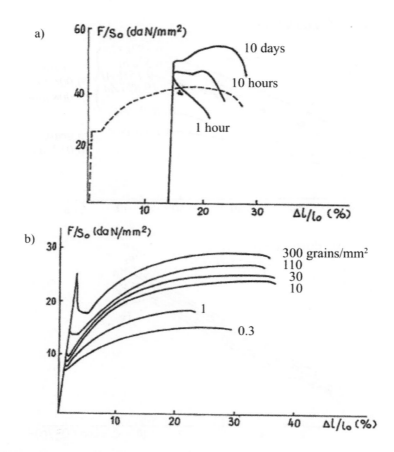

Figure 2.39. *(a) Rough tensile curves of an extra-mild steel pre-hardened by 16%, then age-hardened at room temperature, as a function of the aging time (Bellot). (b) Tensile curves at room temperature of iron with 0.02% carbon, as a function of the average diameter of the grains (Bouchet et al.)*

2.5.17. *Hardening by precipitates*

2.5.17.1. *Consistent precipitates*

EXAMPLE 2.4.–

The zones where the precipitates are consistent with the matrix cause a significant level of hardening, which can be explained by the effects from size and modulus, but above all by the need for them to be "sheared" by the dislocations because of their size and their close proximity. Thus, there is a necessary shearing work per unit length of dislocation.

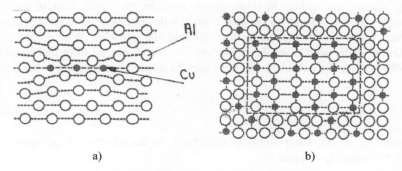

a)   b)

Figure 2.40. *(a) Guinier–Preston area, effects of negative size (atomic radii: Al = 0.143 nm, Cu = 0.128 nm). (b) Consistent precipitate*

2.5.17.2. *Inconsistent precipitates*

EXAMPLE 2.5.–

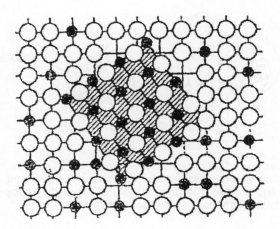

Figure 2.41. *Inconsistent precipitates*

Inconsistent precipitates act as obstacles, which in most cases are impossible for the dislocations to cross, due to their structure that differs greatly from that of the matrix network. This means that the processes for circumventing these obstacles that remain possible are those indicated below (Figure 2.32).

Hardening is created by precipitates with hard dispersoids in close proximity. Orowan indicates that in a ductile matrix made up of particles intervening in the sliding planes, the sliding in these planes requires the application of sufficient stresses for the dislocations to produce a circular development between the particles, in which loops will consequently form around the particles. With regard to the calculation of the sources emitting dislocations, for an escape from the dislocation without shearing the particles, it is necessary to increase stresses $\Delta\tau$ to the amount of:

$$\Delta\tau \simeq \mu\, b/d \qquad\qquad\qquad [2.17]$$

and, with τ_m as the shear stress of the matrix in the absence of precipitate, the overall resistance τ_y is equal to:

$$\tau_y = \tau_m + \mu\, b/d \qquad\qquad\qquad [2.18]$$

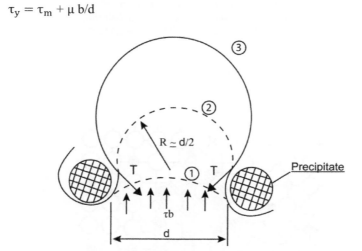

Figure 2.42. *Bypassing of obstacles by dislocation. In ① subcritical position; in ② critical configuration: a semicircle with force τb, d on the segment is compensated precisely by the line voltage T acting on both sides of the precipitate; in ③, the escape phase with the formation of loops around the inclusions*

NOTE.– In general, for the hardening that corresponds to d: distance between particles, and R: radius of the particles, we have:

$$\Delta\tau \simeq \alpha\,\mu\, b/d \text{ with } \alpha \simeq 0.8$$

$$\text{or } \Delta\tau \simeq \beta\,(\mu\, b/d)\, \text{Log}\,(2\,R/b) \text{ with } \beta \simeq 0.2.$$

However, unlike the case of solid solutions and areas (consistent precipitates), an accumulation of loops forms around the particles (residual dislocations after exhausts), which leads to a more significant hardening that increases with deformation.

The influence of thermal activation is negligible here, but, on the other hand, the accumulation of dislocations can increase the embrittlement of the precipitates, leading to a drop in consolidation.

Let us return to the case of consistent precipitates. It is possible for them to be sheared by the dislocations with a force per unit length of the dislocations equal to $\tau_{shearing}$ x b and, for a length d, we obtain $F = \tau_{cis}$ b d. For the work W_{cis} for the supposed spherical particle of radius r, this work is equal to:

$$W_{cis} \simeq F \times 2r \qquad [2.19]$$

By calling γ (I m^{-2}) the fault energy per unit area, we have for the shear on π r^2:

$$W_\gamma = \pi\, r^2\, \gamma \qquad [2.20]$$

where:

$W_{cis} \simeq W_\gamma$ and it gives:

$$\tau_{cis}\ b\ d \times 2r \simeq \pi\, r^2\, \gamma$$

That is:

$$\tau_{cis} \simeq \pi\, r\, \gamma/2\, b\, d \qquad [2.21]$$

In the case of a cubic-type arrangement of the precipitates (edge d), it is possible to define the volume fraction f of precipitates:

$$f = (4/3\ \pi\, r^3)\,/\,d^3 \qquad [2.22]$$

where:

$$\frac{r}{d} = [(3/4\,\pi)f]^{1/3} \qquad [2.23]$$

which, for the shear stress τ, gives:

$$\tau_{cis} \simeq \frac{\pi\gamma}{2\,b}\frac{r}{d} = \frac{\pi\gamma}{2b} \times \left(\frac{3}{4\pi}\right)^{1/3} f^{1/3}$$

$$\tau_{cis} \simeq C\ f^{1/3} \qquad [2.24]$$

In this case, this shows a shear stress of the precipitates that is a function of their volumetric proportion.

NOTE.– Different expressions concern the increase in hardening $\Delta\tau$ as a function of the number of zones found (1/d, with d being the average distance between zones), the occupied volume fraction f, and the energy of "redissolution" ΔU or interface creation γ_s and disorder γ_p. Table 2.11 gives the different expressions of $\Delta\tau_c$. In all cases, the influence of thermal activation is small.

Redissolution effect	$\Delta\tau = \alpha\Delta U/b^2 d$ (flat zones) $f^{1/2}\,\Delta U/b^3$ (spherical areas)
Consistency (or size) effects	$\beta\eta^{3/2}\mu\,(rf\,/\,b)^{1/2}$
Modulus effect	$\beta'\,\Delta\mu\,(b\,/\,d)$
Interface effect	$\gamma_s\,f\,/\,r$
Order effect	$\gamma_p\,f/b$ or $\gamma_p\,f^{1/2}\,/\,b$

Table 2.11. *Hardening due to consistent zones and precipitates* *(r = precipitated radius, $\alpha \simeq \beta \simeq 1$) (according to: Friedel, Fine, Kelly, Nicholson and de Fouquet)*

EXAMPLE 2.6.–

We calculate the overall resistance τ_y of an aluminum alloy with a volumetric proportion f of precipitates inconsistent at a distance of 0.1 μm.

Data:

Re, matrix alone = 70 MPa

μ_{Al} = 26 GPa has a CFC network of = 0.4 nm

We obtain $\Delta\tau$ due to precipitates = $\mu b/d$ with b = a $\sqrt{2}$ /:

b = 0.4 $\sqrt{2}$ /2 = 0.283 nm

and:

$\Delta\tau = 26 \times 10^9 . 0.283 \times 10^{-9}/0.1 \times 10^{-6} = 73.5$ MPa giving, for the alloy:

$$\tau_y = Re/2 + \Delta\tau \simeq 108.5 \text{ MPa}$$

For precipitates Al2Mg in the same matrix of Al, we will calculate the critical distance between consistent precipitates, between the bypass and/or shear, knowing that $\gamma_{\text{precipitates}} = 1.420 \text{ mJm}^{-2}$ $r_{Al} = 0.14$ nm, with the volumetric proportion of the precipitates = 14%.

$$\tau_{\text{precipitous shear}} \simeq \pi\, (r\, \gamma)/2b_{Al}\, d \qquad\qquad [2.25]$$

with:

$$f = \left(4/3\ \pi\ r_{\text{precipitate}}^3\right) / d^3 \qquad\qquad [2.26]$$

It gives:

$$r_p = (3f/4\pi)^{1/3} \cdot d \simeq 0.32\ d$$

And thus the bypass-shearing:

$$(\mu b)_{Al}\, /\, d = \pi\, (r \cdot \gamma)_p\, /\, 2\ b_{Al}\, d$$

$$\mu b/d = \pi\, 0.32\ d\, \gamma/2bd$$

We obtain for d:

$$d = 2\ \mu\, b^2/0.32\ \pi\gamma = 2 \times 26 \times 10^9 \cdot 0.283^2 \times 10^{-18}/0.32 \times \pi \times 1.42$$

$$d = 2.9 \text{ nm}$$

2.5.17.3. *Age hardening alloys*

EXAMPLE 2.7.–

Certain aluminum alloys: The case of Al + Cu giving precipitation in GP zones (Guinier-Preston), θ'' and θ' and θ (Al$_2$Cu).

2.5.17.4. *Stages*

– GP zones in the form of flat disks ($\varnothing \simeq 10$ nm, ep $\simeq 1$ nm, d $\simeq 10$ nm) germinate from the solid solution. They are consistent with the matrix but the sides of the discs, although consistent, show a strong deformation.

– GPs grow to form precipitates θ'', while the other GPs dissolve during the release of the Cu which, by diffusion, is incorporated into the θ'', which grow larger (discs whose faces of r $\simeq$ 100 nm are always consistent with the matrix, the sides of thickness $\simeq$ 10 nm are also consistent but with elastic constraints of consistency, the distance between the θ''. d $\simeq$ 100 nm).

– Some precipitates θ'' dissolve and precipitates θ germinate on the dislocations in the matrix, the Cu that is released causes the phase to grow θ' ($\emptyset \simeq 1$ μm, d $\simeq 1$ μm). The Al_2 Cu is completely inconsistent in the form of a globule.

– Table 2.12 shows the age hardening of certain Al alloys.

Alloy Average composition (%)	State	$R_{e0.2}$ (MPa)	R_2 (MPa)	A (%)	Resistance to fatigue* (MPa at 5 x 10^8 cycles)
2014	O	100	200	20	90
4.4 Cu–0.5 Mg–	T4	290	420	18	140
0.8 Mn–0.8 Si	T6	430	480	12	125
2024	O	100	200	20	90
4.4 Cu–1.5 Mg–	T4	320	460	18	140
0.6 Mn	T6	390	475	10	125
6061	O	55	125	25	50
1.0 Mg–0.6 Si–	T4	150	245	22	90
0.2 Cr–0.3 Cu	T6	275	410	17	100
6070	O	70	145	20	65
0.8 Mg–1.4 Si–	T6	365	400	12	100
0.7 Mn–0.3 Cu					
7005	O	85	200	20	–
4.5 Zn–1.4 Mg–	T6	295	360	13	155
0.12 Cr–0.4 Mn–0.15 Zr					
7075	O	105	230	17	–
5.6 Zn–2.5 Mg–	T6	500	570	11	160
1.6 Cu–0.3 Cr					

* Tested through rotational bending

Table 2.12. *Age hardening of some aluminum alloys and their average mechanical properties*

Example of electron micrography and microdiffraction, 20% Al-Ag, quenched at 530°C and tempered at 250°C for 30 min. The Guinier–Preston zones have disappeared in order for the precipitation of platelets to occur.

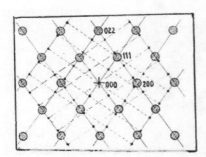

Schematic for interpreting the microdiffraction diagram.

⊘ α (Al) section $(01\bar{1})$
• γ (Ag_2Al) sections $\{01\bar{1}0\}$

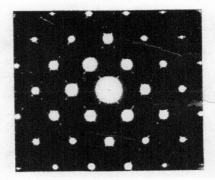

Microdiffraction diagram made on the Al-Ag span. The largest marks form part of the Al matrix, while those with lower intensity belong to two precipitate families in the wafer, respectively (a hexagonal grid with the phase γ – Ag2 Al)

Figure 2.43. *Examples of precipitates (photos: Saulnier, Trillat)*

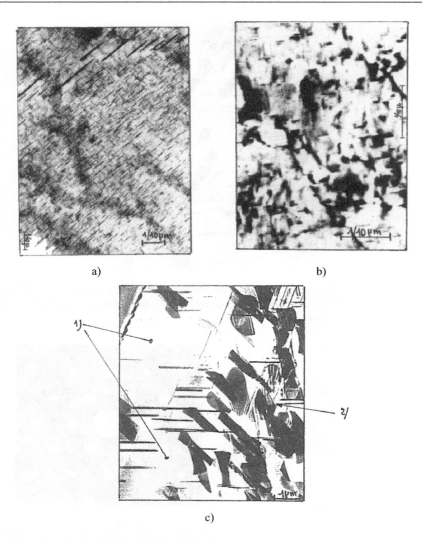

a)   b)

c)

Figure 2.44. *Micrographs of the age hardening of aluminum alloys, showing the presence of zones and precipitates (photos: Saulnier, Trillat)*

COMMENT ON FIGURE 2.44.– *(a) Al-Cu 4%. Quenched, tempered for 21 days at 150°C. Direct electron micrography highlighting the submicroscopic precipitation. (b) Al-Cu 4%: quenched and tempered for 5 h at 160°C. This micrography is analogous to the previous one, but by tilting the preparation, the regions of the matrix deformed by the coherence tensions are revealed (black regions bordering the precipitates). The hardening of the alloy is due to the difficulties experienced by* *the dislocations in passing through these strained regions of the matrix. (c) Al-Ag 20%: quenched at 530°C and tempered for 5 min at 300°C. Two different modes of precipitation exist together: (1) in spherical aggregates of small dimensions (100 Å) known as Guinier–Preston zones and (2) in phase platelets γ - $A_{g2}Al$.*

2.5.17.5. Hardening by ordered precipitates (after Castagné): evolution of mechanical properties

EXAMPLE 2.8.–

This study relates to a nickel-chromium-cobalt-molybdenum austenite hardened by titanium and aluminum, with a composition that is favorable for achieving an ordered coherent precipitation with parameters very close to that of the matrix.

Samples which were subjected to temperatures ranging between 650°C and 1,000°C after hyper-quenching, were observed after a low-level work hardening. It has been observed that the deformation mechanism depends on the size of the precipitates: the dislocations are produced by shearing the nickelides, whose diameter is less than 400 Å, while they bypass the largest precipitates by means of Orowan's mechanism.

A complex nickel-chromium precipitation of titanium and aluminum is obtained by different temperings after hyper-quenching, the precipitates are spherical, homogeneous in size and well distributed within the matrix. They have a very low size factor value, and coherent precipitates are obtained up to diameters of the order of 2,000 To.

Alloy composition is as follows: 55.5 Ni, 19 Cr, 14 Co, 7 Mo, 2.3 Al, 2 Ti, 0.04 C.

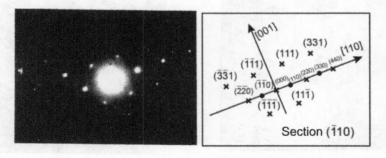

Figure 2.45. *The diffraction diagram shows that the precipitates are ordered (in the case of a sample tempered for 16 h at 725°C) (Castagné 1966). Types of test samples: rolling at 15/100 mm thick, hyper-quenching at 1,180°C in an oil bath, then tempering for 16 h in a vacuum and for 650 < °C < 1,000: 5% traction and thinning in an acetochromic bath*

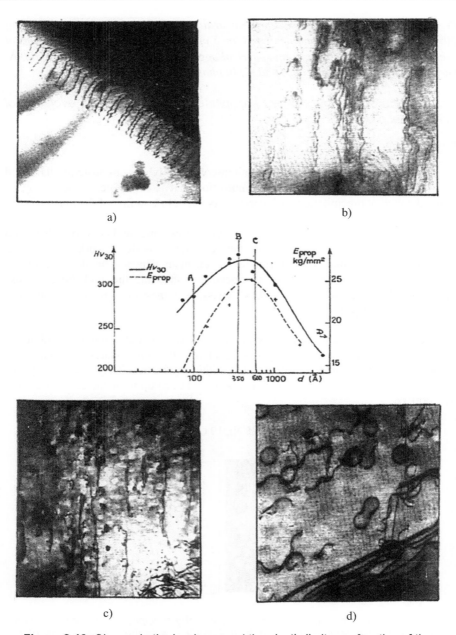

Figure 2.46. *Change in the hardness and the elastic limit as a function of the diameter d of the precipitates, and electron micrographs of dislocation–precipitate interactions*

COMMENT ON FIGURE 2.46.– *(a and b) Shear of the precipitates having d ≃ 100 Å stacks and d ≃ 300 Å oscillating movements of de-placement under electronic bombardment. (c) Presence of loops due to the start of bypassing of the precipitates via Orowan's mechanism and (d) for d ≃ 2,000 Å bypassing with stacks of several stable loops (Castagné et al.).*

Electron microscopy examinations allow us to obtain the histogram of the sizes of the precipitates and their percentages through tempering, and the nature of the dislocation-precipitate interactions, as well as their influences on the hardness (HV) and the elastic shear limit (curves for the diameters of the precipitates and tempering temperatures).

2.5.17.6. Precipitates

Alloy system	Parent phase and lattice(a)	Precipitate phase and lattice(a)	Crystallographic relations (precipitate phase described first)
Ag-Al	Al solid solution; fcc	γ (Ag₂Al); hcp	(0001) // (111), [112̄0] // [11̄0]
	Al solid solution; fcc	γ' (transitional); hcp	(0001) // (111), [112̄0] // [11̄0]
Ag-Cu	Ag solid solution; fcc	Cu solid solution; fcc	Plates // (100); all directions //
	Cu solid solution; fcc	Ag solid solution; fcc	Plates // (111) or (100); all directions //
Ag-Zn	β (βAgZn); bcc	Ag solid solution; fcc	(111) // (110), [11̄0] // [1̄11]
	β (βAgZn); bcc	γ (γAg₃Zn₃); bcc	(100) // (100), [010] // [011]
Al-Cu	Al solid solution; fcc	θ (CuAl₂); bct	Plates // (100); (100) // (100), [011] // [120]
	Al solid solution; fcc	θ' (transitional); tet	(001) // (100), [010] // [011]
Al-Mg	Al solid solution; fcc	β (β-Al₃Mg₂); fcc	Plates first // (110); later probably // (120)
Al-Mg-Si	Al solid solution; fcc	Mg₂Si; fcc	Plates // (100)
Al-Zn	Al solid solution; fcc	Nearly pure Zn; hcp	Plates // (111); (0001) // (111), [112̄0] // <110>
Au-Cu(b)	Au-Cu solid solution; fcc	α"₁ (AuCu I); ord fct	(100) // (100), [010] // [010]
Be-Cu	Cu solid solution; fcc	γ₂ (γ-BeCu); ord bcc	G-P zones // (100); later γ₂ with [100] // [100], [010] // [011]
0.4C-Fe	Austenite (γ-Fe); fcc	Ferrite (α-Fe) (proeutectoid); bcc	(110) // (111), [11̄1] // [11̄0]
0.8C-Fe	Austenite (γ-Fe); fcc	Ferrite in pearlite; bcc	(011) // (001), [1̄00] // [100], [01̄1] // [010]
	Austenite (γ-Fe); fcc	Ferrite in upper bainite; bcc	(110) // (111), [11̄0] // [2̄11]
		Ferrite in lower bainite; bcc	(110) // (111), [11̄1] // [11̄0]
1.3C-Fe	Austenite (γ-Fe); fcc	Cementite (Fe₃C); ortho	Plates not // (111); (001) Fe₃C // to plane of plate
Co-Cu	Cu solid solution; fcc	α-Co solid solution; fcc	Plates // (100); lattice orientation same as parent matrix
Co-Pt(b)	Pt-Co solid solution; fcc	α" (CoPt); ord fct	Plates // (100); all directions //
Cu-Fe	Cu solid solution; fcc	γ-Fe (transitional); fcc	Cubes (100); lattice orientation same as parent matrix
		α-Fe; bcc	Plates // (111); lattice orientation random
Cu-Ni-Co	Cu solid solution; fcc	α-Co solid solution; fcc	Plates // (100); lattice orientation same as parent matrix
Cu-Ni-Fe	Cu solid solution; fcc	α-Fe solid solution; fcc	Plates // (100); lattice orientation same as parent matrix
Cu-Si	Cu solid solution; fcc	β (ζ Cu-Si); hcp	Plates // (111); (0001) // (111), [112̄0] // [11̄0]
Cu-Sn	β phase; bcc	Cu solid solution; fcc	(111) // (110), [11̄0] // [1̄11]
Cu-Zn	β (CuZn); bcc	Cu solid solution; fcc	(111) // (110), [11̄0] // [1̄11]; variable habit; plates or needles // [556]
	β (CuZn); bcc	γ (γ-Cu₅Zn₈); ord bcc	(100) // (100), [010] // [010]
	ε (εCu-Zn); hcp	Zn solid solution; hcp	(101̄4) // (101̄4), [112̄0] // [112̄0]
Fe-N	Ferrite (α-Fe); bcc	γ₁ (Fe₄N); fcc	(112) // (210)
Fe-P	Ferrite (α-Fe); bcc	δ (Fe₃P); bct	Plates // (21,1,4)
Pb-Sb	Pb solid solution; fcc	Sb solid solution; rhom	(001) // (111), [100] // [11̄0]

(a) bcc = body-centered cubic; bct = body-centered tetragonal; fcc = face-centered cubic; hcp = hexagonal, close-packed; ord bcc = ordered body-centered cubic; ord fct = ordered face-centered tetragonal; ortho = orthorhombic; rhom = rhombohedral; tet = tetragonal. (b) Ordering transformation.

Table 2.13. *Crystallographic relationships between precipitates, and constituents for different alloys (according to Metals Handbook, vol. 8, 8th ed., p. 177)*

2.6. Athermal mechanism of the movement of a dislocation

A dislocation of length ℓ will be set in motion due to the effect of a reduced shear stress τ inversely proportional to its length ℓ:

$$\tau = \alpha\, Gb/\ell = \alpha\, Gb\sqrt{\varrho} \qquad\qquad [2.27]$$

In the event that it is not free, it will behave as a source of mobile dislocations (dislocations pinned by other specific dislocations or defects, etc.).

For a medium-sized dislocation network $\bar{\ell}$ in which the segments can be distributed according to their length, we can define a critical split as:

$$\tau_c = \propto Gb/\bar{\ell} = \propto Gb \sqrt{\varrho} \qquad\qquad [2.28]$$

where ϱ is the average density of the dislocation network; if this density is not modified, the variation of τ_c on the basis of the temperature is limited to that of the product Gb. On the other hand, the evolution of ϱ with the strain will be the determining factor of the consolidation.

The first displacements represent a low shear stress with few mobile dislocations (low number of active sources). At the critical threshold, τ_c prevents the emission of numerous loops that travel long distances, in which the emission occurs due to a large number of active sources.

Figure 2.47 illustrates athermal intrinsic processes and the values of $\propto$ occurring in the relationship [2.28] with $\varrho \simeq \ell^{-2}$, ℓ being the average distance between dislocations or the length of the active segments.

In the athermal domain, the sliding mechanisms are blocked by the field of long-distance interactions (>10b), and the domains over which the interactions extend involve relatively large numbers of atoms simultaneously; the overall energy levels are thus much higher than the thermal fluctuations.

For obstacles along the dislocations or at a distance in the order of b or multiple amounts of b, the thermal fluctuations facilitate the passage of the dislocations through these obstacles (activated thermal processes).

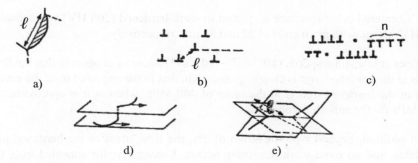

Figure 2.47. *Threshold of $\tau_c \approx \propto Gb\, \varrho^{1/2}$ of athermal intrinsic processes with, for $\propto$ in (a) Frank source of $\approx$ 0.3; (b) remote interactions $\approx$ 0.15; (c) stacks $\approx$ n x 0.15; (d) Jog trail $\approx$ 0.2; (e) attractive junctions $\approx$ 0.3 – 0.2 (according to Otte and Hren 1966)*

2.6.1. *Density ϱ of dislocations and athermal stress*

The athermal threshold stress τ_c [2.28] is a function of the density of the dislocations. Klepaczko et al. (1985) propose a law for the change in the density of dislocations with plastic strain, leading to the following relationship:

$$\varrho = \varrho_o + \frac{M_0}{k_a}\left[1 - \exp(-k_a\gamma)\right] \tag{2.29}$$

where:

- ϱ_o is the initial density of dislocations;

- $M_0\,(\dot{\gamma})$ is the coefficient of the multiplication of the dislocations;

- $k_a\,(\dot{\gamma}, T)$ is the constant of the destruction of the dislocations.

The athermal stress then becomes:

$$\tau_a = \propto Gb\,\sqrt{\varrho_{o+\frac{M_0}{k_a}[1-\exp(-k_a\gamma)]}} \tag{2.30}$$

In compressions, it is given as:

$$\sigma_a = \sqrt{3}\alpha Gb\,\sqrt{\varrho_0 + \frac{M_0}{k_a}\left[1 - \exp(-\sqrt{3}k_a\varepsilon)\right]}$$

As an example, we will look at the study (Leroy et al. 1997a; Canto 1998) on Armco iron stressed in static and dynamic compression ($\dot{\varepsilon} > 10^3 s^{-1}$). The iron $\propto$

with a centered cubic structure is studied in work-hardened (206 HV) and annealed (94 HV) states, with grain sizes of 22 and 31 μm, respectively.

Tests at low strain speeds ($10^{-5}s^{-1} - 10^{-3}s^{-1}$) allow us to observe that the flow stress in the hardened iron is clearly greater than that in the annealed iron; the elastic limit of the hardened iron is in the order of 440 MPa, whereas it is approximately 150 MPa for the annealed iron.

In addition, beyond a plastic strain of 2%, the flow stress of the hardened iron saturates, and no more work hardening occurs. Conversely, for annealed iron, the stress remains constant up to 1% of strain and, beyond that, a consolidation of the material is observed (Figure 2.48).

The static tests at $10^{-5}s^{-1}$ give us the values of the athermal constraints, making it possible to deduce from the relation [2.30] the values of ϱ_0, M_0 and K_a for a value of $\propto = 0.5$ (a value estimated by several authors).

	ϱ_0 in m^{-2}	M_0 (at $10^{-5}s^{-1}$) in m^{-2}	K_a (at $10^{-5}s^{-1}$)
Annealed iron	6.16×10^{13}	1.94×10^{15}	2.0
Hardened iron	6.63×10^{14}	17.0×10^{15}	37.5

Table 2.14. *Initial dislocation densities ϱ_0, coefficient of multiplication M_0 and destruction K_a to $10^{-5}s^{-1}$ for annealed iron and hardened iron (Canto 1998)*

NOTE.– Behavior in dynamics: ($\dot{\varepsilon} > 10^3 \ s^{-1}$).

From the static tests carried out at ambient temperature on samples pre-hardened in dynamic at imposed deformation rates, it can be noted from these curves as follows:

– For annealed iron and work-hardened iron, the static flow stress after a dynamic pre-hardening is lower than the flow stress at the same deformation rate of the deformed sample exclusively in static. These differences show the effect of the strain history. We will highlight this effect from the dynamic–dynamic tests (*jump tests*).

– The annealed iron continues to consolidate statically after an initial dynamic deformation, while the work-hardened iron reaches its maximum flow stress immediately and no longer consolidates during the test.

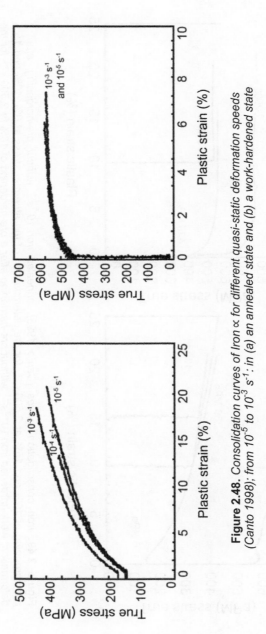

Figure 2.48. Consolidation curves of Iron ∝ for different quasi-static deformation speeds (Canto 1998); from 10^{-5} to 10^{-3} s^{-1}: in (a) an annealed state and (b) a work-hardened state

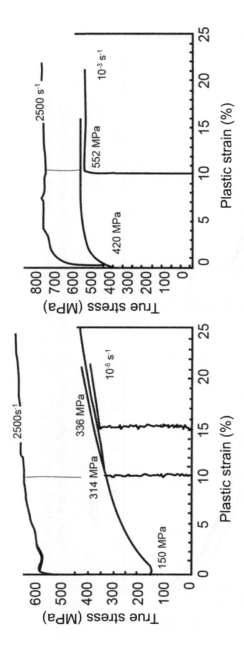

Figure 2.49. *Iron consolidation curves (Canto 1998): annealed iron $10^{-5} s^{-1}$ without pre-hardening, after a pre-hardening of 10% and 15% at 2,500 s^{-1}, for the determination of the athermal stress at 0%, 10% and 15% deformation; work-hardened iron $10^{-3} s^{-1}$ without pre-hardening after a pre-hardening of 10% to 2,500 s^{-1} for the determination of the athermal stress of 0% and 10% deformation*

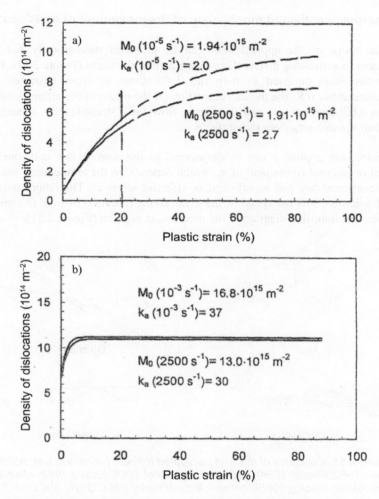

Figure 2.50. *Change in the density of dislocations with deformation and deformation speed; in (a) case of annealed iron at $10^{-5} s^{-1}$ and 2,500 s^{-1} and (b) case of the hardened iron (Canto 1998)*

The values of the stresses at the different rates of deformation in dynamics make it possible to trace the evolution of the total dislocation density with the deformation at $10^{-5}s^{-1}$ and 2,500 s^{-1} (Figure 2.50). It can be seen that there is almost no difference between these two speeds for annealed iron (with $\varepsilon < 20\%$) and for hardened iron.

2.7. Thermally activated mechanism of the movement of a dislocation

If the work of the applied load can be provided mechanically but proves insufficient in surpassing different obstacles at short distances (Figure 2.51), thermal fluctuations must be used to help the dislocations to cross obstacles. For a temperature above 0 K, the thermal activation (at the origin of the atomic vibrations) provides additional energy, and the plastic flow for a stress level τ appears to be lower than the stress τ_0 from the obstacle.

The incision applied τ can be considered as the sum of two components: an athermal or internal component of τ_i, which depends on the temperature solely due to the shear modulus, and an efficient or effective stress τ^*. The latter, assisted by thermal agitation, sets the speed of the dislocations (duration of time to cross local obstacles). A simplified diagram of the mechanism is given (Figure 2.51).

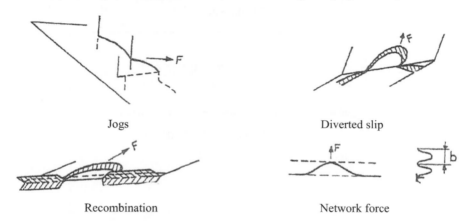

Jogs   Diverted slip

Recombination   Network force

Figure 2.51. *Examples of thermally activated intrinsic processes (ref: notches (McLean 1962; Friedel 1964); deflected slip (Jaoul 1965; Escaig 1968; Alers 1970); recombination (Escaig 1968), network force (Friedel 1964; Guyot and Dorn 1967))*

– If $\tau < \tau_i$ (applied stress < internal stress), the dislocation will not be able to move between the obstacles.

– If $\tau > \tau_i$, after a certain displacement occurs, the dislocation can then enter into position against the obstacle. This exerts a short-range return force. The true stress that takes part in the crossing is $\tau_{\text{efficiency}}$, referred to as τ^*, and equal to $\tau^* = \tau - \tau_i$. It provides part of the energy W^* necessary for the crossing. The additional energy ΔG^* is provided by thermal activation.

– If the temperature (T) is zero, the barrier can only be crossed under a stress of $\tau > \tau_i + \tau_{obstacle}$.

– At a temperature (T), the effective stress τ^* is just enough so that, under the effect of this stress and that of the thermal agitation characteristic of (T), the obstacles are able to be crossed at the desired frequency of the required strain speed.

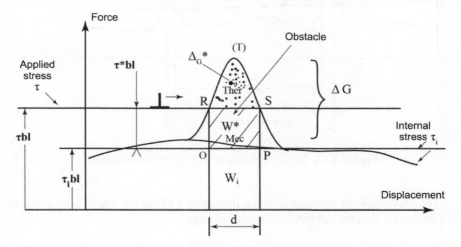

Figure 2.52. *Simplified diagram of the thermally activated mechanism, with the contribution of ΔG^* in overcoming the obstacle for the duration t_1. ΔG (ORTSP area): energy to overcome the obstacle; W^*(ORSP area): mechanical part; ΔG^*(RTS area): thermal assistance part; ℓbd: activation volume*

$$\Delta G = W^* + \Delta G^* \qquad [2.31]$$

When the dislocation moves, its speed is determined by the duration t_1 of the obstacle and the duration t_2 established by the part of its line that has become free to travel the average distance D to the next obstacle (Figure 2.53).

The average speed of the dislocations is thus given by:

$$v \simeq \frac{D}{t_1 + t_2}$$

NOTE.– See additional information on the durations in Appendix A.

In the field where the movement is governed by the thermal activation process, the longest duration controls the strain speed with $t_1 \gg t_2$. We obtain:

$$v \simeq \frac{D}{t_1} \qquad [2.32]$$

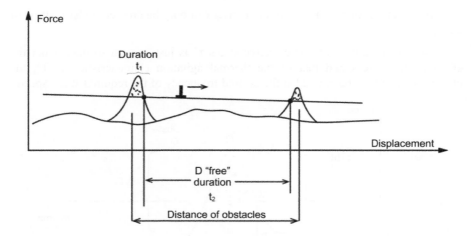

Figure 2.53. *Schematic diagram of the dislocation path*

The inverse of the duration t_1 is the frequency v of the passing of the obstacles; this frequency follows a Boltzmann statistical law:

$$\frac{1}{t_1} = v = v_0 \exp\left(-\frac{\Delta G^*}{kT}\right)$$ [2.33]

where v_0 is the vibration frequency of the dislocation line, ΔG^* is the free activation energy necessary to overcome the obstacle that holds a dislocation in place, k is the Boltzmann constant and T is the absolute temperature.

Knowing that the speed of the slip (or deformation) $\dot{\gamma}$ is equal to (Orowan):

$$\dot{\gamma} = \varrho_m bv$$ [2.34]

where ϱ_m represents the density of the mobile dislocations with a Burgers vector B.

By applying [2.33], we obtain:

$$\dot{\gamma} = \dot{\gamma}_0 \exp\left(-\frac{\Delta G^*}{kT}\right)$$

The activation volume lbd "models" the type of obstacle that is found.

Mechanisms	Activation volume in b^3
Peierls–Nabarro forces	10–100
Intersection of forests	100–10,000
Non-conservative movement of jogs	100–10,000
Diverted slip	10–100
Dislocation climb	1

Table 2.15. *Values of the thermal activation volumes according to Conrad*

As an example, in the case of CC materials, the plastic strain appears to be controlled mainly by the intrinsic resistance of the crystal lattice to plastic shear, according to the Peierls–Nabarro-type mechanism (Conrad 1964). The figure represents the crossing of the Peierls–Nabarro valleys by the double detachment of a dislocation anchored at two points, A and B, distant from l. This is then subject to the effective stress τ^* and, during its displacement, a double dropout of length w and width is formed a^*.

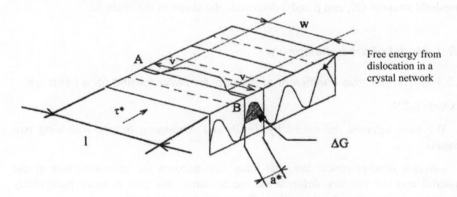

Free energy from dislocation in a crystal network

Figure 2.54. *Scheme for the crossing of valleys from Peierls–Nabarro via double setback*

The vibration frequency of the dislocation line is thus $\nu_0 = \frac{b}{2w}\,\nu_D$, where ν_D represents the vibration of the atoms (Debye frequency 10^{13} Hz). Once the obstacle is crossed, this allows the dislocation to travel an average distance of $s^* = \frac{la^*}{w}$.

Equation [2.33] then becomes:

$$\nu = \nu_D \frac{la*b}{2w^2} \exp\left(-\frac{\Delta G^*}{kT}\right)$$
[2.35]

And taking into account the relationship in [2.34], we obtain:

$$\dot{\gamma} = \dot{\gamma}_0 \exp\left(-\frac{\Delta G(\tau^*)}{kT}\right) = \nu_D \frac{la^*}{2w^2} b^2 \, \varrho_m \exp\left(-\frac{\Delta G^*}{kT}\right)$$
[2.36]

The stress is not explicit in this equation, but it appears in the term for the free work for activation. In the literature, several forms can be found to express $\Delta G(\tau^*)$, the definition of which essentially depends on the shape of the barriers and the structure of the material. However, the relationship given by Kocks et al. (1975), one that is commonly used, gives a good approximation of the free work:

$$\Delta G(\tau^*) = \Delta G_0 \left[1 - \left(\frac{\tau^*}{\tau_0^*}\right)^p\right]^q$$
[2.37]

where ΔG_0 is the total energy of the obstacle, τ_0^* is the effective part of the threshold stress at 0K, and p and q determine the shape of the obstacle.

2.7.1. Model of the behavior law

2.7.1.1. Comparative coefficient of a model for behavior and for a jump test

EXAMPLE 2.9.–

We have selected the model by Zerilli and Armstrong for the following two reasons:

– it is a semi-empirical law that takes into account the microstructure of the material and the primary deformation mechanisms; this part is more particularly interesting to us within the thermal mechanism that is activated;

– it is a model that distinguishes the behavior of CFCs and the CCs, particularly with a study on iron α.

For CFCs:

$$\sigma = \Delta\sigma_G + B_1\varepsilon^{1/2} \exp[(1 - \beta_0 + \beta_1 \ln \dot{\varepsilon})T] + k_\varepsilon \ell^{-1/2}$$
[2.38]

For the CC:

$$\sigma = \Delta\sigma_G + B_0 \exp[(-\beta_0 + \beta_1 \ln\dot{\varepsilon})T] + k_0\varepsilon^n + k_\varepsilon \ell^{-1/2}$$
[2.39]

with:

– T: temperature;

– $\dot{\varepsilon}$: the deformation speed;

– ε: deformation;

– ℓ: the polycrystal grain diameter;

– $\Delta\sigma_G$, B_0, B_1, β_0, β_1, K_0, n, k_ε: experimental constants based on the analysis of the mechanisms of plastic strain in these two crystal lattices.

The expression [2.39] can be separated into two components: an athermal component $\sigma_{ath}(z)$, and a thermal component σ_{th}, which can be compared, respectively, to the dynamic internal constraint σ_i or the static athermal stress σ_{ath} and the effective stress σ^*:

$$\sigma_{ath(z)} = \Delta\sigma_G + K_0\varepsilon^n + k_\varepsilon\ell^{-1/2}$$

– $\Delta\sigma_G$ reflects the influence of the solute and the initial dislocation density on the stress field;

– $K_0\varepsilon^n$ is the term that translates the consolidation during plasticization;

– $k_\varepsilon\ell^{-1/2}$ translates the stress concentrations in the grain boundaries necessary for the transmission of the plastic flow between the grains.

$$\sigma_{th} = B_0 \exp\left[(-\beta_0 + \beta_1 \ln \dot{\varepsilon})T\right]$$

In the case of Armco iron, the authors determined the constants mainly from the experimental results of Johnson and Cook.

$\Delta\sigma_G$(MP$_a$)	K_0(MP$_a$)	n	k_ε (MP$_a$.mm$^{1/2}$)	B_0 MP$_a$	β_0 (K^{-1})	β_1 (K^{-1})
0	266.0	0.289	22.0	1,033	0.00698	0.000415

Table 2.16. *Model constants determined by Zerilli and Armstrong for Armco iron (Zerilli et al. 1987)*

2.7.1.2. Comparison of the model in a jump test

EXAMPLE 2.10.–

Application of the Zerilli and Armstrong model for annealed and work-hardened Fe , a study of the dynamic behavior of a CC at dynamic-dynamic jump (jump test). Study of the change in the different coefficients of the model ("structural memory") (Leroy et al. 1984, 1997; Canto 1998).

Figure 2.55 shows the experimental behavior curves for annealed iron at strain rates of 2,500 s^{-1} and 1,600 s^{-1}, and for work-hardened iron at the deformation rate of 1,600 s^{-1}. The same figure shows the theoretical curves obtained using the model of Zerilli and Armstrong, with the constants above, T = 293 K, ℓ = 22 μm for the work-hardened iron, and ℓ = 31 μm for the annealed iron.

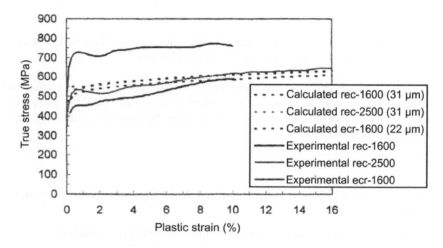

Figure 2.55. *Experimental behavior curves calculated using the work of Zerilli and Armstrong in the case of annealed iron for strain speeds of 2,500 s^{-1} and 1,600 s^{-1}, and work-hardened iron for 1,600 s^{-1} (Leroy et al. 1997; Canto 1998)*

It can be observed that the curves calculated are very close to each other, while experimentally, the differences between the flow stresses for different initial loads or states are greater. Now we will analyze the athermal stress and the thermal stress independently.

By using the different instantaneous responses to the dynamic–dynamic jumps, we can readjust these coefficients for these three loads.

NOTE.– For iron α, a comparison of our experimental results with the model of Zerilli and Armstrong's law of behavior shows that the constants used do not allow the history of the strain to be taken into account. Dynamic–dynamic jumps make it possible to determine the evolution of different coefficients of the model (Canto 1998).

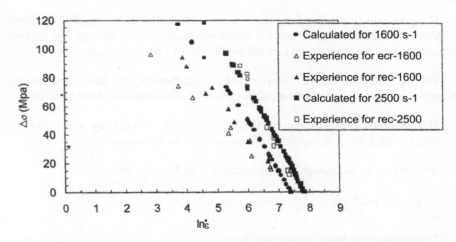

Figure 2.56. *Comparison between the experimental values at the dynamic–dynamic jumps of $\Delta\sigma^*$ and the calculated values of $\Delta\sigma_{th}$, deduced from the works of Zerilli and Armstrong. The case of annealed iron at $\dot{\varepsilon}_1$ is equal to 1,600 s^{-1} and 2,500 s^{-1}, and work-hardened iron at $\dot{\varepsilon}_1$ is equal to 1,600 s^{-1} (Leroy et al. 1997; Canto 1998)*

	B_0 (MPa)	β_0 (K^{-1})	β_1 (K^{-1})
Rec-2500	2,240	0.01030	0.000476
Rec-1600	2,240	0.01270	0.000590
Ecr-1600	2,240	0.01385	0.000633

Table 2.17. *Coefficients B_0, β_0 and β_1 from the Zerilli and Armstrong model, determined from the dynamic–dynamic jumps (Canto 1998)*

2.8. The viscous friction mechanism

When the stress levels are high enough to mechanically force the dislocations to cross all the barriers without the assistance of thermal fluctuations, t_1 becomes negligible when compared with t_2, the viscous damping effects become predominant. This domain is characterized by a linear increase in the stress when the deformation speed increases.

The frictional force $F = \tau_B^* b$, acting per unit length on the dislocation in its sliding plane, causes a displacement of the movable dislocation at a limit speed v_ℓ such that:

$$v_\ell = \frac{\tau_B^* \cdot b}{B}$$

τ_B^* is the effective stress from the viscous friction, that is to say, the difference between the applied stress τ and the stress necessary to pass through the barriers without thermal assistance, and B is the coefficient of viscous friction.

This coefficient is the sum of the contribution of five mechanisms: thermo-elastic, sonic and electronic diffusions, and sonic and electronic viscosities.

The theoretical estimation of the coefficient is limited by the condition $v_\ell < c/3$ (with c being the speed of the sound wave in the material).

By delaying v_ℓ in Orowan's relationship, $\dot{\gamma} = \varrho_m b v_\ell$, we infer:

$$\tau = \tau_0 + \frac{B}{\varrho_m b^2}\,\dot{\gamma}$$

This relationship leads to empirical laws:

$$\sigma = A + \eta\dot{\varepsilon}$$

The equation of motion of the dislocation is related to the viscous damping coefficient B by the relation:

$$m\frac{dv}{dt} + Bv = \tau\,\left|\vec{b}\right| \qquad\qquad [2.40]$$

where:

– m is the mass per unit length of the dislocation;

– v is the speed of the dislocation;

– τ is the reduced split;

– $\vec{b}$ is the Burgers vector;

– B is the viscous damping coefficient for interactions of dislocations-phonons and dislocations-electrons.

The coefficient B depends on the temperature. The solution of the equation [2.40] is:

$$v = v_\ell\,(1 - e^{-(B/m)t}) \qquad\qquad [2.41]$$

where $v_\ell = \tau b/B$ represents the limiting speed of the dislocation.

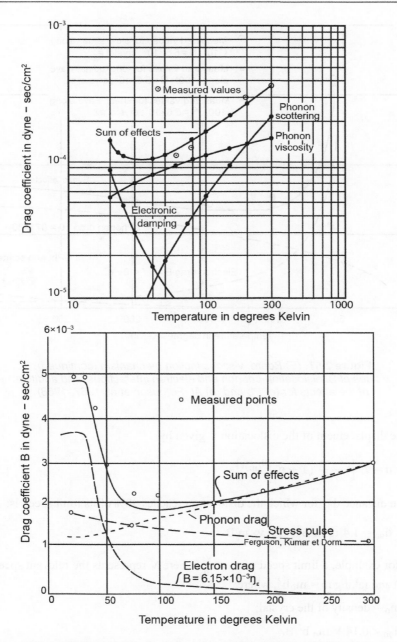

Figure 2.57. *(A) Board: viscous friction mechanism. Values of B, dislocation-phonon and electron interactions, and addition of the effects for (a) Pb and (b) Al (Ferguson et al. 1967, 1968)*

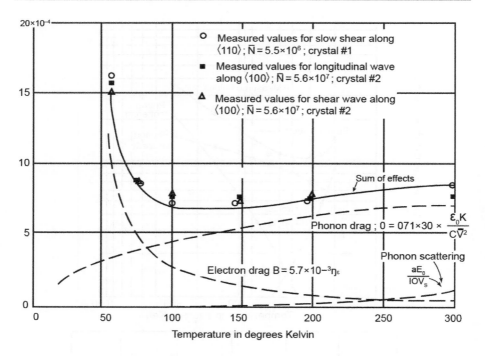

Figure 2.57. *(B) Board: viscous friction mechanism (continued). Values of B, dislocation-phonon and electron interactions, and addition of the effects for (a) Pb and (b) Al (Ferguson et al. 1967, 1968)*

The displacement of the dislocation is given by:

$$d = v_\ell \left[t - \frac{m}{B} \left(1 - e^{-(B/m)t} \right) \right]$$ [2.42]

The distance d_{90} for which the dislocation reaches 90% of its limit speed is:

$$d_{90} = 1.4 \, \tau b \, m/B^2 = 1.4 \, v_\ell \, m/B$$

with, for example, a limit speed of $v_\ell < \frac{V}{10}$ (where V represents the relevant speed of sound) and taking $m = m_0 b^2$, where:

– m_0 = density of the crystal;

– $d_{90} < 0.14 \, V \, m_0 \, b^2/B$.

The internal friction measurements (Alers and Thompson 1961), ultrasonic attenuation (Hikata and Elbaum 1967; Mason 1968) and displacement velocity of dislocations by application of stress impulses (Hauser et al. 1961; Kumar and Kumble 1969), make it possible to determine an average value of B: $B \sim 10^{-4}$ dyne.s.cm^{-2}.

Or $d_{90} < 8b$.

The dislocation reaches its limit speed v_ℓ after a course of less than 8b by taking $v_\ell < \dfrac{v}{10}$. For lead, Parameswaran and Weertman (1969) give $v_\ell \sim 5.5 \times 10^4 \dfrac{\text{cm}}{\text{s}}$ at 78 K, that is, two-thirds of the speed of sound. In this case, d_{90} is superior and $d_{90_{Pb,78K}} < 50b$.

We can therefore consider that, in the case of large slips, the dislocation reaches its limit speed after a few meshes of the atomic grid, and equation [2.41] reduces to:

$$v_\ell = \frac{\tau b}{B} \qquad\qquad [2.43]$$

Thus, the damping term B plays a key role in the value of the deformation in the case of highly dynamic stresses. Effectively, the plastic deformation rate d ε_p is directly connected to B in the case of a polycrystal by:

$$d\,\varepsilon_p = \frac{1}{m} \sum \frac{N_{S,M}\, b^2 \tau}{B} \qquad\qquad [2.44]$$

This sum extends to all the sliding mechanisms of the crystals, where m is the Taylor coefficient and $N_{S,M}$ is the density of mobile dislocation segments.

The density $N_{S,M}$ of mobile dislocation segments is difficult to evaluate. Some authors give it as a constant, while Ferguson indicates that:

$$N_{S,M_{300K}} / N_{S,M_{70K}} \simeq 3$$

Ferguson et al. (1967, 1968) show that the plastic strain rate changes at a linear rate with stress, which is to say that:

$$\tau - \tau_0 = \eta \cdot \dot{\varepsilon} \qquad\qquad [2.45]$$

where η is a temperature-dependent constant and τ_0 a constant.

Since $v_\ell = (\tau - \tau_0)\, b/B$ and $\dot{\varepsilon} = \varrho_m\, b\, v_\ell$, together with [2.45], we obtain:

$$B = \eta\, \varrho_m\, b^2 \qquad\qquad [2.46]$$

It can be noted that, when the stress applied is high, the second large slip threshold intervenes in particular, and the applied force must allow the dislocation to cross the dislocations which pierce through the slip plane.

2.8.1. *Influence of viscosity η of the medium on the viscous damping coefficient B*[1]

By calculating the total loss of energy W_V (with a moving dislocation screw), we can obtain B_V, and it can in fact be shown that:

$$W_{vis} = \frac{b^2 v^2}{8 \pi a_{0V}^2} \eta$$

where a_0 is the radius of the core of the dislocation and η the viscosity of the medium, thus:

$$B_{vis} = \frac{b^2 \eta}{8 \pi a_{0V}^2} \tag{2.47}$$

The same is obtained for an edge dislocation:

$$B_{coin} \sim \frac{3}{4} \frac{b^2 \eta}{8 \pi (1-v)^2 a_{0C}^2} \tag{2.48}$$

In general, a_0 is considered to be equal to 3/4 b by many authors, and the phonic viscosity η_p is equal to:

$$\eta_p = \frac{E_0 K_p}{C_{vp} \overline{V}} \tag{2.49}$$

where:

– $E_0 = \hbar \omega$ is the total thermal energy, the sum of the energies of the possible modes of vibration of the metal;

– K_p is the thermal conductivity;

– C_{VP} is the specific heat due to phonons;

– $\overline{V}$ is the average speed of Debye.

---

1 See supplemental material in Appendix A.

Pippard et al. (1955) give the following equation for the value of the electronic viscosity η_e:

$$\eta_e = 9 \times 10^{11} \, h^2 \, (3 \, \pi^2 \, N)^{2/3} \, 5 \, e^2 \, \rho \qquad [2.50]$$

where N represents the number of electrons per cm^3 and ρ the electrical resistivity of the medium.

Another cause of damping is due to the phenomenon of diffusion of phonons, which is linked to the movement of dislocations. Leibfried considers that the depreciation B_D, in this case, is equal to:

$$B_D = \frac{a E_0}{10 \, V_S} \qquad [2.51]$$

where a is the network parameter and V_S is the speed of the shear waves.

The value of B_D is low compared to those of B_P and B_e.

3

Dynamic Flows for Monocrystals and Polycrystals

– Technique used: internal electromagnetic shearing.

– Interest:

- internal shock of a mechanical nature without contact;

- high strain speed $\dot{\gamma} \simeq 10^4$ s^{-1};

- visualization of crystallographic flows;

- influence of phonic and electronic viscoplasticities.

The test samples, in the form of monocrystal and polycrystal rings, are cut by an eroding machine. The dimensions of the rings are as follows: ϕ_{ext} = 20 and 30 mm; ϕ_{int} = 13 and 20 mm; thickness = 0.5 and 1 mm.

We control these dimensions and measure the changes in them after being strained by electromagnetic forces, using a Trioptics machine that allows for a precise measurement. The rings previously drawn with circles (Figure 3.1) are placed in a temperature-varying cryostat, which allows them to be cooled from 293K to 1K, to then be deformed using generators, which produce magnetic pulses that allow us to obtain shear speeds in the order of 10^4 s^{-1}, where the dynamic strains are observed by an ultra-fast camera.

ADDITIONAL MATERIALS.– Metallography: article by Maurice Leroy, transmitted by Louis Néel in: Leroy, M. (1970a). Viscoplasticité des métaux cubiques à faces centrées. *C. R. Acad. Sc. Paris*, Series C, 270, 899–902.

See supplemental materials in Appendix D.2, volume 1: www.iste.co.uk/leroy/annexes.pdf.

3.1. Type of monocristal and polycrystal dynamic shear test samples (anisotropy and isotropy at high speeds)

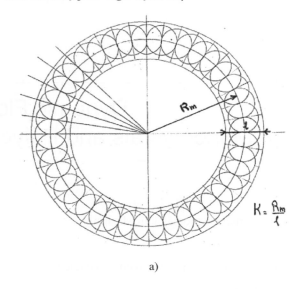

a)

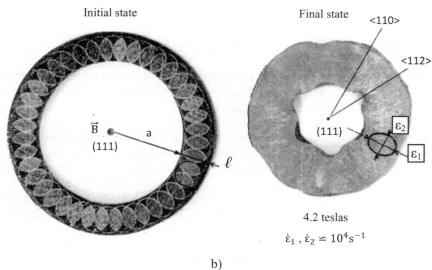

b)

Figure 3.1. *Example of dynamic electromagnetic crystallographic flow by shearing $\dot{\varepsilon} \simeq 10^4 s^{-1}$. The case of a planar copper crystal (111) stressed at 4.2 T at ambient temperature (Leroy 1997). (a) Test piece engraved with circles Ø 4 mm; Ø ext = 30, Ø int = 20, thickness = 1 mm; K = Rm/l = 2.5. (b) Flow anisotropy of the crystal (111)*

3.2. The tensor of the shock stresses

$$K = Rm/\ell$$

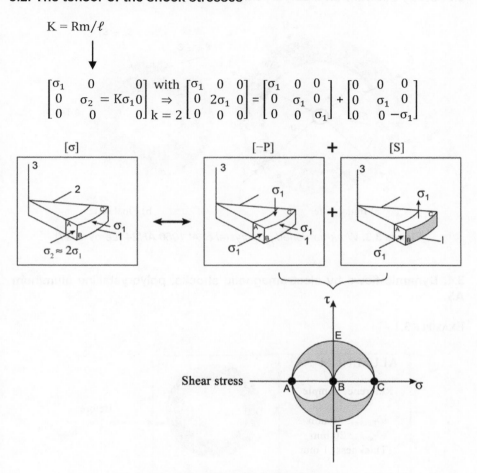

$$\begin{bmatrix} \sigma_1 & 0 & 0 \\ 0 & \sigma_2 & = K\sigma_1 0 \\ 0 & 0 & 0 \end{bmatrix} \text{ with } \begin{array}{c} \begin{bmatrix} \sigma_1 & 0 & 0 \\ 0 & 2\sigma_1 & 0 \\ 0 & 0 & 0 \end{bmatrix} \\ \Rightarrow \\ k = 2 \end{array} = \begin{bmatrix} \sigma_1 & 0 & 0 \\ 0 & \sigma_1 & 0 \\ 0 & 0 & \sigma_1 \end{bmatrix} + \begin{bmatrix} 0 & 0 & 0 \\ 0 & \sigma_1 & 0 \\ 0 & 0 & -\sigma_1 \end{bmatrix}$$

Figure 3.2. *Stress on test element*

Compression: $k = 2$, pressure p and shear by $|s|$.

3.3. Study of strain on a polycrystal

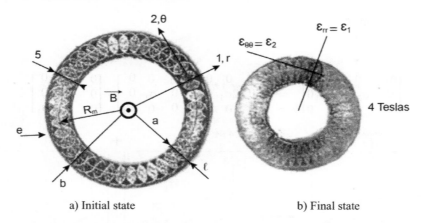

a) Initial state   b) Final state

Figure 3.3. *Work-hardened polycrystal of Al 1050 AH24 (Leroy 1997)*

3.4. Dynamic flows by electromagnetic shocks, polycrystalline aluminum A5

EXAMPLE 3.1.–

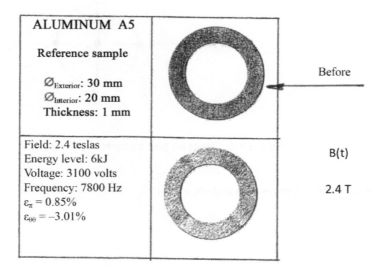

ALUMINUM A5		
Reference sample		Before
$\varnothing_{\text{Exterior}}$: **30 mm** $\varnothing_{\text{Interior}}$: **20 mm** **Thickness: 1 mm**		
Field: 2.4 teslas Energy level: 6kJ Voltage: 3100 volts Frequency: 7800 Hz $\varepsilon_\pi = 0.85\%$ $\varepsilon_{\theta\theta} = -3.01\%$		B(t) 2.4 T

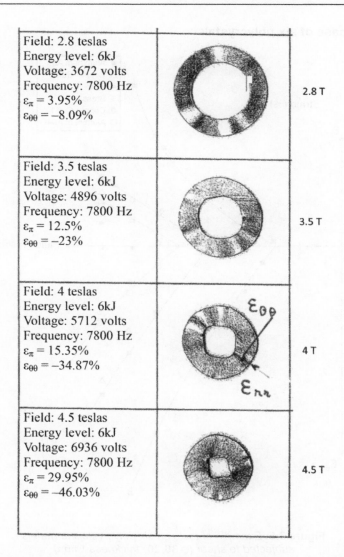

Field: 2.8 teslas Energy level: 6kJ Voltage: 3672 volts Frequency: 7800 Hz $\varepsilon_\pi = 3.95\%$ $\varepsilon_{\theta\theta} = -8.09\%$	2.8 T
Field: 3.5 teslas Energy level: 6kJ Voltage: 4896 volts Frequency: 7800 Hz $\varepsilon_\pi = 12.5\%$ $\varepsilon_{\theta\theta} = -23\%$	3.5 T
Field: 4 teslas Energy level: 6kJ Voltage: 5712 volts Frequency: 7800 Hz $\varepsilon_\pi = 15.35\%$ $\varepsilon_{\theta\theta} = -34.87\%$	4 T
Field: 4.5 teslas Energy level: 6kJ Voltage: 6936 volts Frequency: 7800 Hz $\varepsilon_\pi = 29.95\%$ $\varepsilon_{\theta\theta} = -46.03\%$	4.5 T

Figure 3.4. *Example of dynamic strain by electronic action on aluminum A5 test samples with values according to the intensities of the magnetic induction pulses, K = Rm/l = 12.5/5 = 2.5 (Leroy 1997)*

$$\varepsilon_{rr} = \frac{\text{width}_{(final)} - \text{width}_{(initial)}}{\text{width}_{(initial)}} \quad \varepsilon_{\theta\theta} = \frac{\Phi\,\text{mean}_{(final)} - \Phi\,\text{moyen}_{(initial)}}{\Phi\,\text{mean}_{(initial)}}$$

3.5. The case of six polycrystals

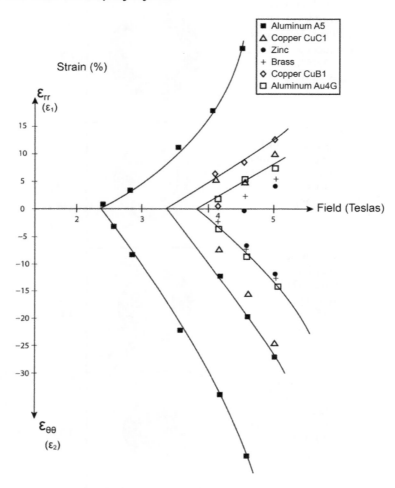

Figure 3.5. *Example of tests on six polycrystalline metals subjected to shear (ϕ 30, 20, thickness 1 mm)*

COMMENT ON FIGURE 3.5.– *Experimental curves of the change in strain as a function of the intensities of the applied fields. Tests on a 6 KJ generator ($C = 180 \ \mu F$), $f = 7.8 \ kHz$, peak of pulses at $T/4 = 32 \ \mu s$. For 1 tesla = 1 Wb/m^2 and for $\mu_r = 1$, $H = 0.7958 \times 10^{-6} \ A/m$, with $e \geq \delta : P_0 = \frac{\mu_0}{2} H_0^2$, that is, $\simeq 0.4 \ MPa$ for 1 tesla $K = 2.5$ (geometric factor of the test samples) (Leroy 1997).*

Materials used	Elastic limit (MPa)	Resistance (Ω m. 10^6)
Aluminum A5	110	0.027
Aluminum Au4G	268	0.027
Copper CuB1	170	0.0175
Copper CuC1	230	0.0175
Brass	172	–
Bronze	480	–
Monel	182	–
Zinc	109	0.06

Table 3.1. *Elastic limit in "static" tension of the various polycrystals subjected to dynamic stress by magnetoforming*

Metal	Field threshold (teslas) for radial deformations	Field threshold (teslas) For axial deformations
Aluminum A5	2.5	2.4
Aluminum Au4G	3.85	3.75
Copper CuB1	3.5	3.6
Copper CuC1	3.5	3.6
Brass UZ 36	3.75	4
Pure zinc	3.65	4.25

Table 3.2. *Plasticity thresholds of the various materials subjected to dynamic stress by magnetoforming*

3.6. The case of monocrystals

Figure 3.7: example of the radial dynamic flow of monocrystals, ε_{rr} obtained from planes with cuts of (100), (110), and (111) made on a single crystal of aluminum, annular discs of the same initial dimension, all stressed with the same impulse B(t) of 4.7 T.

Figure 3.8 shows the monocrystalline sections of the stereographic triangle stressed by dynamic shear, in order to produce the mapping of the cristallographic flows for two types of area axes; plaster models from the maps are given.

Figures 3.9, 3.12, 3.13: comparison of anisotropic flows of CFC (Al, Cu) (plaster models of the axes of crystallographic zones.

3.6.1. *Flow of a crystal*

EXAMPLE 3.2.–

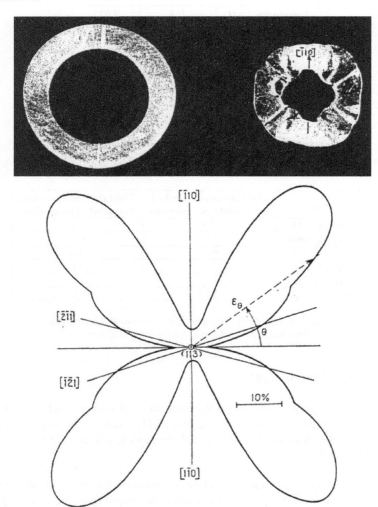

Figure 3.6. *Anisotropy of the flow ε_{rr} (ε_θ) of a monocrystallic aluminum ring (plane of the ring (113)). Value of the dynamic stress 10^8 dynes/cm^2, duration of the stress near 15 μs by electromagnetic shear. The photo shows the crystal before and after the shock, which gives a radial crystallographic flow ε_{rr} on the cutting plane (113) as given by θ (Leroy 1972)*

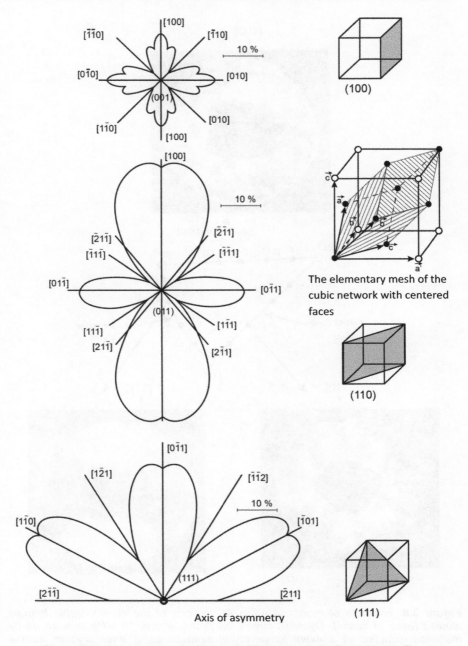

Figure 3.7. *Comparison of anisotropic "flows" ε_{rr} (ε_{θ}) of monocrystalline cuts of aluminum stressed at 4.7 T (Leroy 1975)*

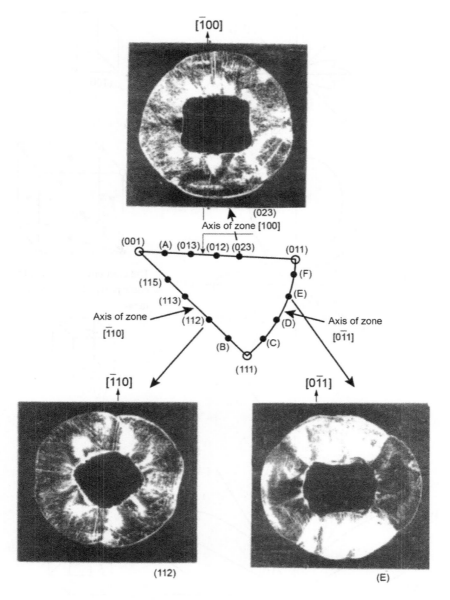

Figure 3.8. *Example of monocrystalline Al sections of the stereographic triangle studied (axes of zones). Dynamic stress, electronic action, 10 MPa in ≃ 15 μs by magnetic induction at ambient temperature, example using three crystals from a stereographic triangle (Leroy 1972)*

NOTE.– Comparisons of anisotropic flows of the two axes of zones [110] and [100] for Al and Cu, respectively, appear in Figures 3.9, 3.12 and 3.17 (Al), and 3.13 and 3.20 (Cu).

3.7. Models for CFCs

The monocrystalline sections of the stereographic triangle studied are the following:

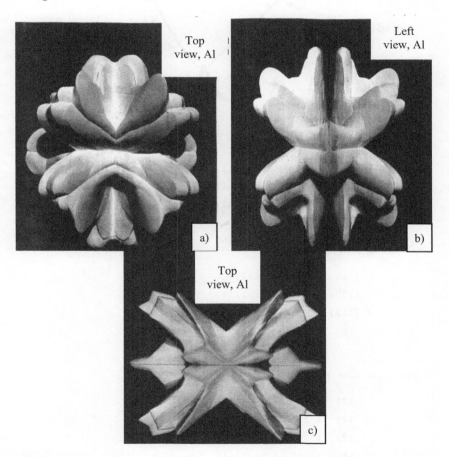

Figure 3.9. *Representative model of the dynamic flows of mono-crystalline aluminum sections, zone axis [110], stress by a magnetic pulse of 4.7 teslas, that is, 10 MPa in ≃ 15 μs, T = 293K. (Leroy 1972). Aluminum (zone axis of type [110]), model in (a), (b) and (c)*

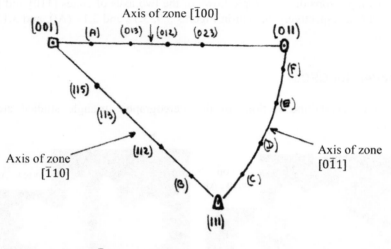

Axis of zone [$\bar{1}00$] :

        (023) : 11°15' of (011)

        (012) : 18°30' of (011)

        (013) : 26°30' of (011)

        (A)   : 36°30' of (011)

Axis of zone [$\bar{1}10$] :

        (115) : 39° of (111)

        (113) : 29°30' of (111)

        (112) : 19°30' of (111)

        (B)   : 9° of (111)

Axis of zone [$0\bar{1}1$] :

        (C) : 9° of (111)

        (D) : 19°30' of (111)

        (E) : 29°30' of (111)

        (F) : 31°30' of (111)

Figure 3.10. *Monocrystal cuts*

3.7.1. *Viscoplastic dynamic flows of crystals: the case of aluminum* ($\dot{\varepsilon} \simeq 10^4 \ s^{-1}$)

EXAMPLE 3.3.–

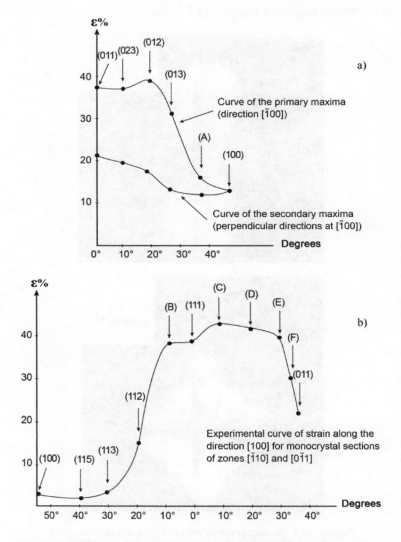

Figure 3.11. *Crystallographic flows by electromagnetic shock applied to annular Al crystals at 293K (planar disks, $\emptyset_e$ = 20 mm, $\emptyset_i$ = 14 mm, p: 0.5 mm, K = 2.8) in (a) zone axis [$\bar{1}00$] ; b) [$\bar{1}10$] and [$0\bar{1}1$] (Leroy 1972)*

NOTE.– The measurement of overall deformations ε_t of these three single-crystal cuts (100), (110) and (111) gives:

– For copper: $\varepsilon_{t(111)} = 3.5\ \varepsilon_{t(110)} = 6.8\ \varepsilon_{t(100)}$;

– For aluminum: $\varepsilon_{t(111)} = 2\ \varepsilon_{t(110)} = 12.5\ \varepsilon_{t(100)}$.

Front view, Al

Top view, Al

Figure 3.12. *Representative model (monocrystalline cuts) (Leroy 1972). Aluminum (zone axis of type [100])*

5,6 Wb/m²

a) Stereographic representation of the experimental strain of planar monocrystalline samples of CFC (copper). Zone axis [$\bar{1}$10] in front view

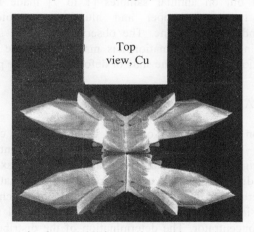

b) Zone axis of type (110) in top view (Chiem and Leroy 1975)

Figure 3.13. *Stereographic representation of the experimental strain of planar monocrystalline samples of CFC (copper). Zone axis [$\bar{1}$10] seen from above (Chiem and Leroy 1975). Copper (zone axis of type [110])*

3.8. Dynamics of flows shown using an ultra-fast camera

Abstract. Mono and/or polycrystalline ring-shaped specimens are submitted to magnetic induction of several Teslas in a field concentrator. Biaxial stresses are induced inside the specimen, due to radial compression, and lead to the dynamic deformation of metal at strain rates of about 103 s-1 to 104 s-1 and for plastic deformation up to 40%. The test ring geometry allows us to get stress and strain states involved and, in particular, to deform the monocrystal (111) in shearing according to <110>. During the magnetic impulse period, the evolution of deformation has been observed with a high speed camera and then the paths of stress and strain states have been plotted for different field intensities. We compare those paths and energy levels of CFC copper mono- and polycrystals with aluminum.

1) INTRODUCTION

The objective of this study is to analyze the behavior at high speed and at high strain rates of metals subjected to the action of intense and pulsed magnetic fields. The electromagnetic pressures applied in the test sample generate intense shocks without mechanical contact, and their effects on the materials are analyzed. The tests were carried out on annular samples [1 to 7] made from copper and aluminum. Two types of copper and aluminum samples were tested: monocrystalline and polycrystalline. The observation of the samples by high-speed camera during their deformation has made it possible to evaluate their speeds of deformation, as well as the state of deformation in the plane of the ring.

2) EXPERIMENTATION

Magnetic induction pulses B are obtained by discharging capacitances in a high-strength coil. This coil is equipped with a field concentrator of a useful diameter Φ and height h. The annular plane samples with an axis of revolution z are subjected to radial loading. The observation of the deformation is done using an ultra-fast camera. The images that are obtained are synchronized with the recordings of the intensities of the current through the coil and those of the induction in the concentrator. The determination of the distributions of induced current densities J(Am-2), of the volumetric forces (Nm-3), is made by finite elements using the ANSOFT axisymmetric code. The metals that were studied are the following:

– polycrystals of work-hardened aluminum 1050AH24 (electrical conductivity $\gamma = 3.6 \times 107$ Ω-1m-1, static yield strength $\sigma E = 114$ MPa);

– annealed and work-hardened copper polycrystals, CuAl ($\gamma = 5.8 \times 107$ Ω-1m-1, $\sigma E = 88.7$ and 261 MPa);

– aluminum monocrystals with z axes <100>, <110>, <111> and copper <111> of respective thermal mass capacities CpCu = 385 and CpAl = 890 J/kg°C.

Figure 3.14. *Electromagnetic shocks. Observation of the dynamic behavior of metals (Leroy et al. 1997)*

ADDITIONAL MATERIALS.– The text of Figure 3.14 is taken from: Leroy*, M., Nicolazo*, C., Louvigné**, P.F., Thomas**, T. (1997b). Dynamic behavior of metals to magnetic field pulses. *Colloque C3, EURODYMAT 5th International Conference on Mechanical and Physical Behaviour of Materials under Dynamic Loading, J. Phys. IV*, Supplement to *Journal de physique III*, August 1997. C3-121-C3-126.

*Institut universitaire de technologie de Nantes – 3, rue du Maréchal Joffre, BP. 34103, 44041 Nantes Cedex 1, France.

**DGA/DCE, Centre de recherches et d'études d'Arcueil – 16bis, avenue Prieur de la Côte d'Or, 94114 Arcueil, France.

See supplement in Appendix D.1, *J. Phys. IV*, volume 1: www.iste.co.uk/leroy/ annexes.pdf.

3.8.1. *Dynamics of shear shocks*

Equivalent stress $\bar{\sigma}$ at the first induction peak B for t = T/4 = 30μs, with: $\alpha = 64.15 \; 10^{-9}$ for B (tesla), f (kHz), $\gamma(\Omega^{-1}m^{-1})$, R_m and e in mm, K = $R_{m/l}$ and for frequencies between 5 and 15 kHz.

An example of trajectories of deformation observed by ultra-fast camera is given for aluminum and copper.

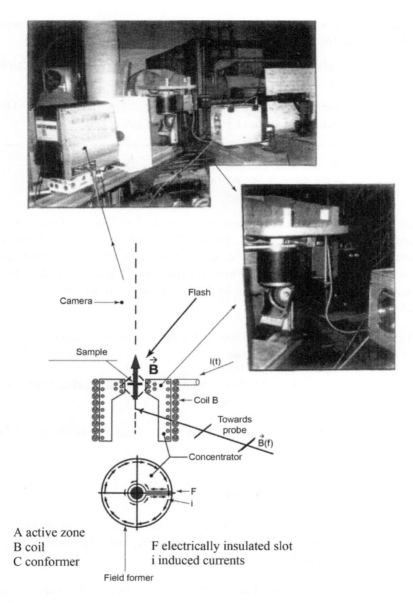

A active zone
B coil
C conformer

F electrically insulated slot
i induced currents

Figure 3.15. *Preparation of magnetic pulse tests. Study using ultra-fast camera, imaging every 5 µs, eight shots, 5 teslas in the field concentrator, 18 Kj generator, deformation in about 20 µs, strain speed of 10,000/s. Work done in collaboration with CTA/DGA, Arcueil (Leroy et al. 1997)*

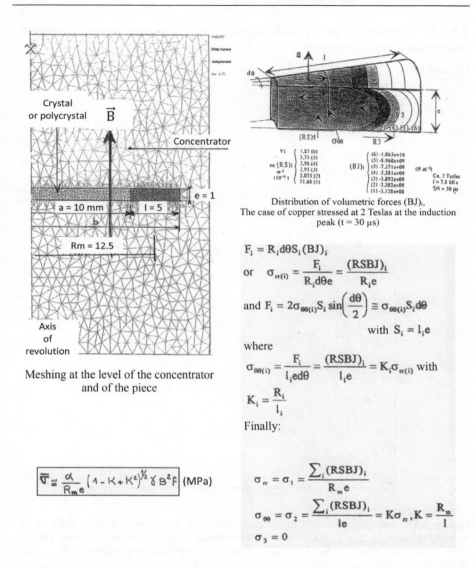

$$F_i = R_i d\theta S_i (BJ)_i$$

or $\quad \sigma_{rr(i)} = \dfrac{F_i}{R_i d\theta e} = \dfrac{(RSBJ)_i}{R_i e}$

and $\quad F_i = 2\sigma_{\theta\theta(i)} S_i \sin\left(\dfrac{d\theta}{2}\right) \cong \sigma_{\theta\theta(i)} S_i d\theta$

with $\quad S_i = l_i e$

where

$$\sigma_{\theta\theta(i)} = \dfrac{F_i}{l_i e d\theta} = \dfrac{(RSBJ)_i}{l_i e} = K_i \sigma_{rr(i)} \text{ with}$$

$$K_i = \dfrac{R_i}{l_i}$$

Finally:

$$\sigma_{rr} = \sigma_1 = \dfrac{\sum_i (RSBJ)_i}{R_m e}$$

$$\sigma_{\theta\theta} = \sigma_2 = \dfrac{\sum_i (RSBJ)_i}{le} = K\sigma_{rr}, K = \dfrac{R_m}{l}$$

$$\sigma_3 = 0$$

Figure 3.16. *Electromagnetic forces in the test piece. Distribution of the values of the internal stresses causing the shearing as a result of an induction pulse B (t). Distribution of the volumetric forces (BJ)i in the case of copper stressed at 2 teslas*

3.8.1.1. *Trajectories of crystallographic flows; dynamic shear; aluminum polycrystals and monocrystals*

EXAMPLE 3.4.–

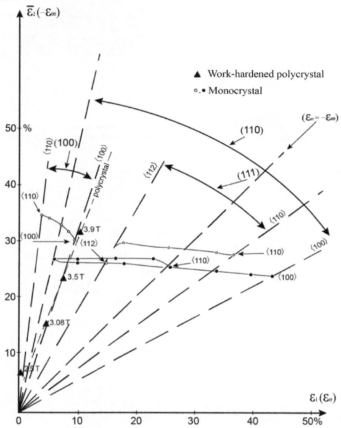

Figure 3.17. *Anisotropy of deformations ε_1 and ε_2 of aluminum crystals (100), (110) and (111), shocks at 4.7 teslas and deformations of aluminum polycrystals (1050) AH24 work-hardened under individual shocks of variable inductions (2.5, ... 3.9 teslas) (Leroy 1997)*

NOTE.– ε_1 (ε_{rr}) is the radial flow rate, and ε_2 ($\varepsilon_{\theta\theta}$) is the circumferential rate (negative). The interval between each point on the flow curves corresponds to 5 µs ($\dot{\varepsilon} \simeq 10^4 \text{s}^{-1}$).

3.8.1.2. *Electronic micrography of crystallographic flow by dynamic shear*

EXAMPLE 3.5.–

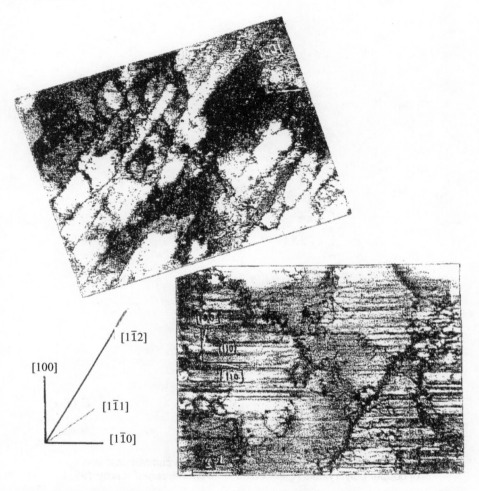

Figure 3.18. *Micrographs of an Al crystal (110), stressed by an electromagnetic pulse of 4.7 teslas of 15 μs, image on a thin plate in the zone of maximum isolation [100], including the presence of dislocation cells with walls of [112] and, during the observation at the MET, movement of the [110] dislocations (Leroy 1972)*

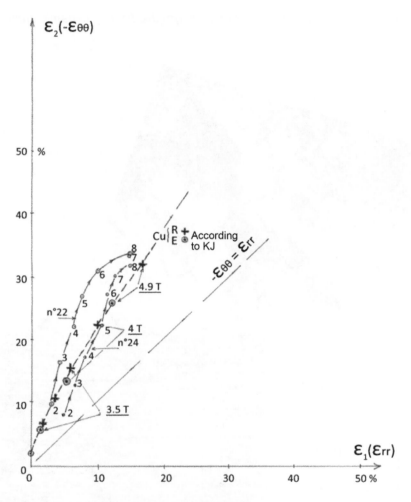

Figure 3.19. *Dynamics of radial flow ε_1 and circumferential flow ε_2 of copper polycrystals R: annealed, E: work-hardened (Leroy 1997)*

3.8.1.3. *Copper monocrystals and polycrystals*

EXAMPLE 3.6.–

The observations using an ultra-fast camera of flows by induction shocks $B_{(t)} = B_0\, e^{-t/\tau} \sin \omega t$ show that the deformations occur only at the first pulse $0 < t \le T/2$, after a delay of about 20 μs for the flow thresholds of the order of 2 teslas. After reaching this threshold, the deformation rates reach 10^4 s^{-1}, with rates for ε_1 (ε_{rr}) and ε_2 ($\varepsilon_{\theta\theta}$) along the straight lines for the entire duration of the flows, which is carried out in 30 µs in volumes $\simeq 400$ mm^3 of the test piece by the volumetric forces (BJ)$_i$ (N·m^{-3}).

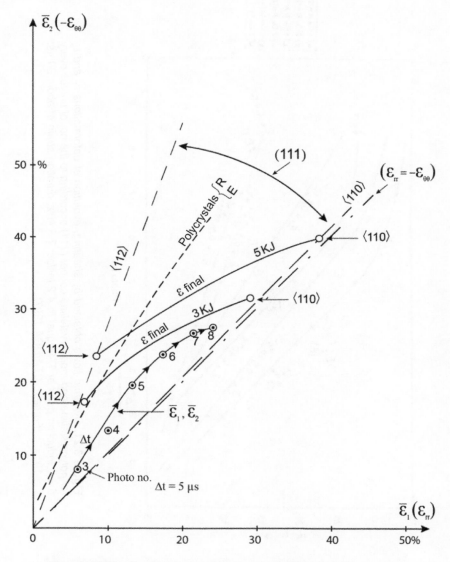

Figure 3.20. *The case of copper monocrystals (111) (Leroy 1997)*

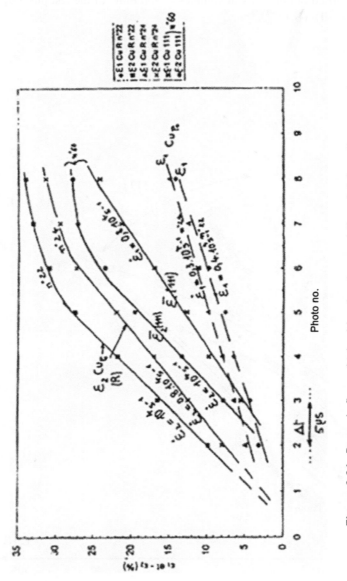

Figure 3.21. *Dynamic flows obtained by shearing, evolution of deformations ε_1 and ε_2 for polycrystalline and monocrystalline copper (111) according to the no. of photos every 5 μs. Induction peak at T/4 equal to 4.2 teslas; 7.8 kHz; duration strain of about 30 μs; $\dot{\varepsilon} \cong 10^4 \ s^{-1}$ (Forcem, CTA/DGA 1997)*

3.9. Viscoplasticity

In the field of flow velocity $\dot{\gamma} \geq 10^4 \text{ s}^{-1}$, a viscoplasticity appears after a domain that is thermally activated at lower speeds.

The stresses will force the dislocations to cross all obstacles without any intervention by thermal fluctuations, and we obtain a linear increase in the stresses with the deformation speeds.

The influence of viscous friction on the speed of dislocations in crystals is studied by measuring the deformations of single crystals of Al, Pb, Zn and on polycrystals of Al and Cu. The interactions of dislocations-phonons and dislocations-electrons form the origin of the viscous friction. We study the influence of temperature on the viscous friction coefficient experimentally from 4K to 300K to distinguish the various interaction models. At the same time, we measure the anisotropic viscoplastic flow in the crystals so as to evaluate the influence of the different viscous frictions.

Figure 3.22. *Low temperature test bench, 6 kJ generator of electrical impulses, equipped with a compression coil and cryostat type A*

Figure 3.23. *Glass enclosures of the polycrystals, allowing for the "transparency" of action of electromagnetic pulses for tests up to 1.8K (Leroy 1997)*

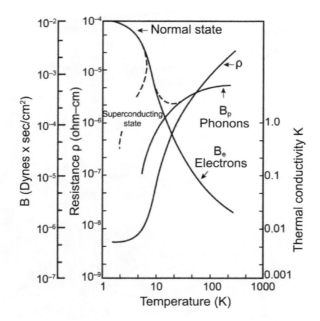

Figure 3.24. *Damping coefficient B for lead in normal and superconducting states (Mason 1966). Note the presence of a sloping variation between 15K and 40K due to the sum of the electronic and phonic viscosities, and the decrease of B in the superconducting state*

3.9.1. *Influence of phonic and electronic viscous friction: the case of lead*

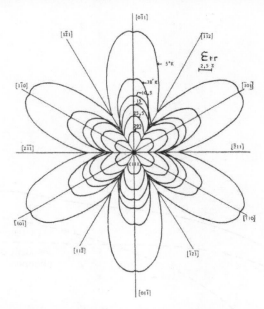

Figure 3.25. *Experimental strain on the basis of the temperature of lead monocrystals, annular axis sections (111), shocks by magnetic field in cryostat, ambient tests at 5 K (superconductivity of Pb). The electromagnetic pulses are of the same intensity, and there is a significant change in the flows ε_{rr} at very low temperatures (order of to 5 K = 6 x ε_{rr} at ambient temperature and 3.5 times for 78 K), with a pulse of 10 MPa (Leroy 1972)*

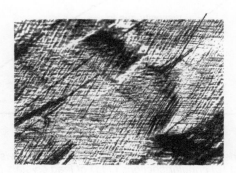

Figure 3.26. *Micrographic aspect of the plane (111). Lead crystal deformed at 5K by a magnetic pulse of 4.7 Wb m^{-2} (Leroy 1972)*

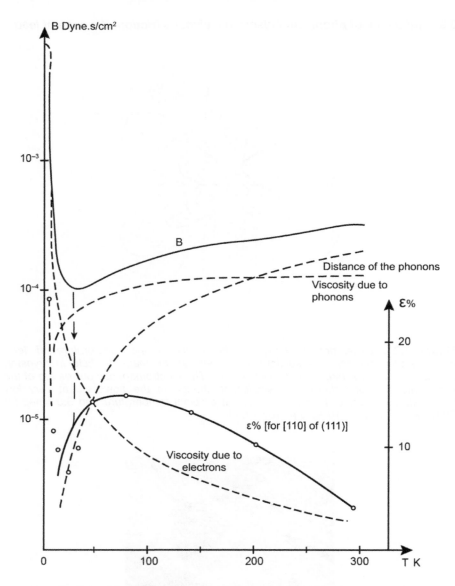

Figure 3.27. *Theoretical and experimental curves of the viscoplastic flow according to [110] annular sections (111) of lead crystals (Leroy 1972)*

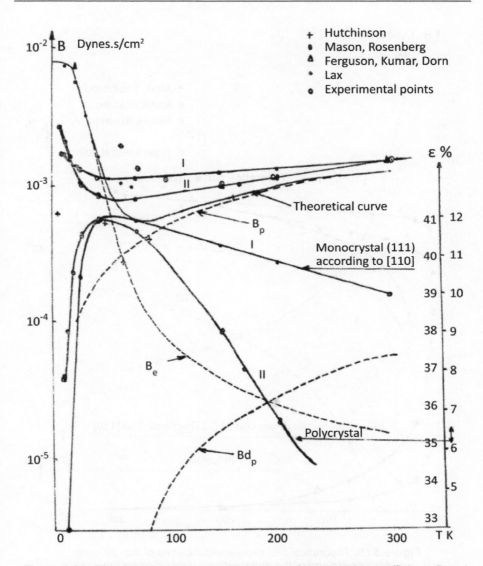

Figure 3.28. *Theoretical and experimental curves of the damping coefficients B and of the deformation ε depending on the temperature of the aluminum. The case of monocrystals according to [110] of sections (111) and polycrystals (Leroy 1972)*

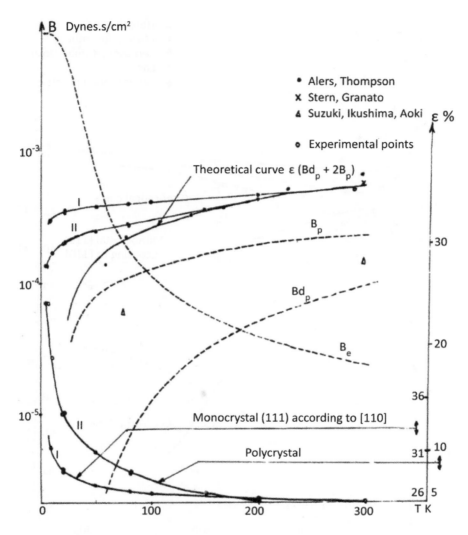

Figure 3.29. *Theoretical and experimental curves of the damping coefficients and of the deformation ε depending on the temperature. The case of copper (Leroy 1972)*

3.9.2. *Viscoplasticity of metal crystals: influence of viscous friction*

Figures 3.30–3.33 are taken from Métallographie, *C. R. Acad. Sc.*, series C (October 1971).

ADDITIONAL MATERIALS.– Metallography: article by Leroy and Offret, transmitted by Louis Néel in: Leroy, M. and Offret, S.C. (1971). Viscoplasticité de cristaux métalliques. Influence du frottement visqueux sur la vitesse des dislocations. *C. R. Acad. Sc. Paris*, Series C, 273, 1049–1052.

The influence of viscous friction on the speed of dislocations is studied by measuring the deformation of metal monocrystals subjected to intense pulse magnetic fields. The experimental determination of the viscous coefficient of friction between 4.2K and 300K allows us to critically examine the various dislocation-phonon and dislocation-electron interaction models.

In the deformation of metals at high speed ($\dot{\varepsilon} > 10^3$ s^{-1}), it can be considered that the strain depends mainly on the viscous damping B of the dislocations. The length, referred to as the path d_{90} for which the dislocation reaches 90% of its limit speed v_1, is equal to:

(1) $$d_{90} = 1.4 v_1 \frac{m}{B}$$

With

(2) $$v_1 = \frac{\tau b}{B}$$

where m is the mass of the dislocation per unit length and τ is the stress exerted.

The measurements of internal friction ([1]) and ultrasonic attenuation [([2]), ([3])] imply that B will have a value of about 10^{-4} dyn·s·cm^{-2} at room temperature. If we assume that $V/10 < v_1 < 2$ V/3 where V is the speed of sound, we calculate that

(3) $$8.|\vec{b}| < d_{90} < 50 |\vec{b}|$$

Thus, the dislocation reaches its limit speed at the end of a path that measures the length of a few atoms.

Considering that the speed of dislocations is v_i in the case of large slides, we can write that the speed of plastic deformation $\dot{\varepsilon}_p$ is a direct function of B and the density N_M mobile dislocations

(4) $$\varepsilon_p = \frac{N_M b^2 \tau}{B}$$

The damping coefficient B depends on the interactions of dislocations-phonons and dislocations-electrons and on the diffusion of the phonons. As a result, it varies depending on the temperature. We can calculate B by evaluating the total loss of energy due to the movement of a dislocation.

Figure 3.30. *Viscoplasticity of metallic crystals (Leroy and Offret 1971)*

If η is the viscosity of the medium, we can observe that for the screw and wedge dislocations, we obtain, respectively:

$$B_v = \frac{b^2\eta}{8\pi a_0^2} \quad \text{and} \quad B_c \simeq \frac{3}{4}\frac{b^2\eta}{8\pi (1-v)^2 a_0^2}$$

where a_0 is the radius of the core of the dislocation and v is Poisson's coefficient. The viscosity Ti is the sum of two terms, the phonic viscosity

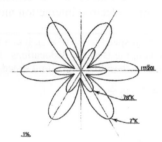

Figure 1. Comparative experimental curves of the viscoplastic flow $_\varepsilon$ in the basal plane (0001) of zinc crystals on the basis of the temperature. Value of the inductor field: 4.7 Wb·m^{-2}.

η_p and the electronic viscosity η_c. We demonstrate [(4), (5)] that

(5) $\eta_p = \frac{E_o K_p}{C_{vp}\bar{V}}$

where K_p is the thermal conductivity, C_{vp} is the specific heat due to the phonons, E_o is the total thermal energy, V is the average speed of Debye, and

$$\eta_c = 9 \times 10^{11}.h^2(3\pi^2N)^{\frac{3}{2}}5c^2\rho$$

where N is the number of electrons per cubic centimeter and p is the electrical resistivity of the metal. In order to experimentally demonstrate the change of (4) with temperature, we subjected planar and annular monocrystalline test samples to loads through intense and pulsed magnetic fields and measured the resulting viscoplastic flow (G). The magnetic field of 4.7 Wb·m^{-2} has been pulsed by 50 µs between 293 and 5 K. The test pieces were made of lead, zinc, aluminum and copper.

Figure 3.31. *Viscoplasticity of metallic crystals (Leroy and Offret 1971)*

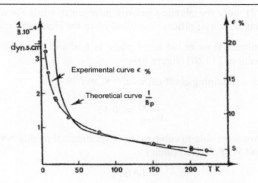

Figure 2. Experimental curve of the viscoplastic flow $_\varepsilon$ and theoretical curve of the phonic damping coefficient according to the temperature for the case of copper.

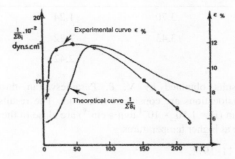

Figure 3. Experimental curve of the viscoplastic flow $_\varepsilon$ and theoretical curve of the electronic and phonic damping coefficients according to the temperature for the case of aluminum.

The measurements of the plastic strain according to the directions (e.g. figure 1) have made it possible to establish the viscoplasticity curves and to highlight the respective contributions of the electronic and phonic viscosities and to compare our experimental values with those given theoretically by Mason ([2]).

Figure 3.32. *Viscoplasticity of metallic crystals (Leroy and Offret 1971)*

For copper (figure 2), only the phonic viscosity intervenes, while for aluminum, the strain peak around 40 K indicates a significant contribution to the electronic viscosity (figure 3).

For zinc, the viscoplastic flow in the basal plane is low and the deformation maxima are obtained for the directions $[11\bar{2}0]$ (figure 1).

The ratio of the viscous damping coefficients at 7 K and 78 K is

$$\frac{B_{7K}}{B_{78K}} = 0.43$$

In the case of lead, we were able to compare our experimental results with those of V. R. Parameswaran ([4]) and W. P. Mason (table).

TABLE

Comparison of the damping coefficients B for lead (10^{-4} dyn·s·cm^{-2}).

	293 K	78 K	4.2 K
W.P. Mason	3.70	1.24	1.50
V.R. Parameswaran	3.43	2.22	1.52
M. Leroy	–	2.42	1.5

The experimental results obtained by V. P. Parameswaran through measuring the displacement of the dislocations are consistent with ours. The results from Mason in the superconducting domain (B = 1.50_ × 10^{-4} dyn·s cm^{-9}) are close to the experimental values, but the same is not true at higher temperatures.

(*) Session of October 11, 1971.

(1) G. A. ALERS and D. O. THOMPSON, J. *Appl. Phys.*, 32, 1961, p. 283.

(2) W. P. MASON, in *Dislocations Dynamics*, 1968.

(3) A. HIKATA and C. ELBAUM, *Phys. Rev. Letters*, 18, 1967, p. 750.

(4) V. P. PARAMESWARAN and J. WEERTMAN, *Scripta Metalurgica*, 3, 1969, pp. 477–480.

(5) W. G. FERGUSON, A. KUMAR, J. E. DORN, *Acta Met.*, 16, 1968, p. 1189.

(6) M. LEROY, *Comptes rendus*, 270, Series C, 1970, p. 899.

Figure 3.33. *Viscoplasticity of metallic crystals (Leroy and Offret 1971)*

3.10. References for viscoplasticity

ADDITIONAL MATERIALS.– Metallography: article by Maurice Leroy, transmitted by Louis Néel: in Leroy, M. (1970a). Viscoplasticité des métaux cubiques à faces centrées. *C. R. Acad. Sc. Paris*, Series C, 270, 899–902.

Article by Leroy and Offret, transmitted by Néel in: Leroy, M. and Offret, S.C. (1971). Viscoplasticité de cristaux métalliques. Influence du frottement visqueux sur la vitesse des dislocations. *C. R. Acad. Sc. Paris*, Series C, 273, 1049–1052.

Physics of solids: article by Chiem, Pouliquen and Leroy, transmitted by Néel in: Chiem, C.Y. and Leroy, M. (1975). Glissement {110} <110> dans les monocristaux de cuivre et d'aluminium déformés par des champs magnétiques. *C. R. Acad. Sc. Paris*, Series B, 235, 282.

Article by Chiem, Renaud and Leroy, presented by Néel in: Leroy, M. (1975). Déformation par champs magnétiques de coupes monocristallines de cuivre et d'aluminium.. *C. R. Acad. Sc. Paris*, Series B, 179, 280.

NOTE.– See supporting information materials in Appendices D.1 and D.2, volume 1: www.iste.co.uk/leroy/rheology1.zip.

Limits to Static and Dynamic Formability

4.1. Plastic instability

4.1.1. *Necking*

Between the yield strength and the maximum load point, the stress–strain curve has a parabolic shape: in this work-hardening, this is the point where "necking" begins. In the zone where this occurs, the test sample is no longer in simple tension, but is subjected to a state of stress along three axes. The average tensile stress $\sigma = F/S$, where S is the section of the necking zone, is no longer identified using the general stress $\overline{\sigma}$ of the material. According to the equivalence rule of mechanical tests, $\overline{\sigma}$ is equal to the value that the tensile stress would have in the necking zone if it remained under simple tension, which is not the case. Bridgman's formula relates $\overline{\sigma}$ to σ:

$$\sigma = \overline{\sigma} \left(1 + \frac{2R}{a}\right) \ln \left(1 + \frac{a}{2R}\right)$$

where a is the radius of the minimum section of the necking zone, and R the radius of the curvature of this zone.

We analyze the instability, noting that if a force F is applied, then the section must transfer this load. If a section A deforms slightly more than the others (see Figure 4.2), its surface is smaller, and therefore the stress is higher there.

If the work hardening has increased the flow stress sufficiently, the reconstructed section can still transmit F:

Transmitted force $F = A\sigma$

– If $A\sigma$ increases with strain, the sample is stable.

– If $A\sigma$ decreases, it is unstable, and thus necking occurs.

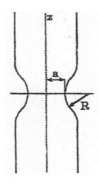

Figure 4.1. *Necking zone*

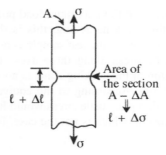

Figure 4.2. *Necking section*

The critical condition of necking is:

$$A\sigma = F = \text{const.}$$

where:

$$A d\sigma + \sigma dA = 0$$

or also:

$$\frac{d\sigma}{\sigma} = -\frac{dA}{A}$$

The conservation of the volume in plastic strain gives:

$$-\frac{dA}{A} = \frac{dl}{l} = d\varepsilon$$

Thus, $\frac{d\sigma}{\sigma} = of\varepsilon$ or also:

Necking point $\frac{d\sigma}{d\varepsilon} = \sigma$

4.1.2. *Work hardening coefficient n*

At the necking point, $\frac{d\sigma}{d\varepsilon} = \sigma$ on the rational curve, and for $\sigma = \sigma_0 + K\varepsilon^n$ of this curve:

$$\frac{d\sigma}{d\varepsilon} = nK\varepsilon^{n-1} \text{ where } n = \varepsilon$$

4.1.3. *Instabilities studies (including speed effect $\dot{\varepsilon}$)*

Nominal strain ε_n: $\varepsilon_n = \frac{l - l_0}{l_0}$ and we will call (sections) $Z = \frac{S_0 - S}{S_0} = 1 - \frac{S}{S_0}$:

$$\hookrightarrow \frac{S}{S_0} = 1 - Z$$

ε: true (rational):

$$\varepsilon = \int_{l_0}^{l} \frac{dl}{l} = l_n \frac{l}{l_0} = l_n (1 + \varepsilon_n) \qquad [4.1]$$

Volume conservation (plasticity):

$$lS = l_0 S_0 \qquad [4.2]$$

where:

$$\frac{l}{l_0} = \frac{S_0}{S} = 1 + \varepsilon_n = \frac{1}{1 - Z} = e^{\varepsilon} \qquad [4.3]$$

and:

$$\frac{dl}{l} = \frac{d\varepsilon_n}{1 + \varepsilon_n} = -\frac{dS}{S} = \frac{dZ}{1 - Z} = d\varepsilon \qquad [4.4]$$

Beginning of necking (plastic instability, early damage) and:

$$\begin{cases} \dfrac{d\sigma}{d\varepsilon} = \sigma \text{ (or nominally } \dfrac{d\sigma_n}{d\varepsilon_n} = 0) \\[2mm] n = \varepsilon \end{cases} \qquad [4.5]$$

While:

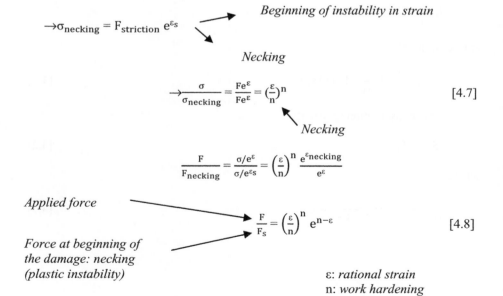

$$FS_0 = \sigma s \rightarrow \sigma = F\frac{S_0}{S} \text{ with } [4.3] \rightarrow \sigma = Fe^{\varepsilon} \qquad [4.6]$$

Initial section

Tensile strength on test sample/unit of section

during necking (force $= F_{striction}$).

Beginning of instability in strain

$$\rightarrow \sigma_{necking} = F_{striction}\, e^{\varepsilon s}$$

Necking

$$\rightarrow \frac{\sigma}{\sigma_{necking}} = \frac{Fe^{\varepsilon}}{Fe^{\varepsilon}} = \left(\frac{\varepsilon}{n}\right)^n \qquad [4.7]$$

Necking

$$\frac{F}{F_{necking}} = \frac{\sigma/e^{\varepsilon}}{\sigma/e^{\varepsilon s}} = \left(\frac{\varepsilon}{n}\right)^n \frac{e^{\varepsilon necking}}{e^{\varepsilon}}$$

Applied force

Force at beginning of the damage: necking (plastic instability)

$$\frac{F}{F_s} = \left(\frac{\varepsilon}{n}\right)^n e^{n-\varepsilon} \qquad [4.8]$$

ε: *rational strain*
n: *work hardening coefficient*

4.1.4. *Role of the strain speed*

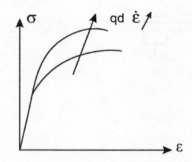

Figure 4.3. *Influence of $\dot\varepsilon$*

We usually have $\sigma\nearrow$ when $\dot\varepsilon\nearrow$:

$$\sigma_{plasticity} = K'\varepsilon^n \dot\varepsilon^m \qquad [4.9]$$

with m: coefficient (which depends on the material).

This is the sensitivity of the stress to the rate of strain or in force:

$$m = dLog^\sigma/dLog\dot\varepsilon$$

$$F = K'\varepsilon^n \dot\varepsilon^m e^{-\varepsilon} \qquad [4.10]$$

The maximum of F (i.e. during necking at the beginning of damage) is:

$$\frac{dF}{d\varepsilon} = 0 \text{ that is to say, for } \varepsilon = \varepsilon_{necking} = n \qquad [4.11]$$

For a tensile test at a set constant speed:

$$\varepsilon_n = \frac{l-l_0}{l_0} = \frac{\Delta l}{l_0} \rightarrow \text{elongation speed of } \Delta l = const.$$

with [4.4]:

$$d\varepsilon = \frac{d(\varepsilon_n)}{1+\varepsilon_n}$$

$\uparrow$

true

True strain speed:

$$\dot{\varepsilon} = \frac{d\varepsilon}{dt} = \frac{d(\varepsilon_n)}{1+\varepsilon_n}/dt = \underbrace{\frac{1}{1+\varepsilon_n}\underbrace{\frac{d(\varepsilon_n)}{dt}}_{\dot{\varepsilon}_n}}$$

with [4.3]:

$$1 + \varepsilon_n = e^{\varepsilon}$$

hence:

$$\dot{\varepsilon} = \dot{\varepsilon}_n \cdot e^{-\varepsilon} \tag{4.12}$$

$$d\dot{\varepsilon} = -\dot{\varepsilon}_n \, e^{-\varepsilon} \, d\varepsilon = \dot{\varepsilon} \, d\varepsilon$$

where:

$$\frac{d\dot{\varepsilon}}{d\varepsilon} = \dot{\varepsilon} \tag{4.13}$$

at instability (always at the set elongation speed):

(see [4.11]) $V\frac{dF}{d\varepsilon} = 0 \; \frac{1}{F}\frac{dF}{d\varepsilon} = 0$ and $F = K'\varepsilon^n\dot{\varepsilon}\,e^{-\varepsilon}$ (see [4.10]) and $\frac{d\dot{\varepsilon}}{d\varepsilon} = \dot{\varepsilon}$ (see [4.13])

$$\longrightarrow \frac{n}{\varepsilon} + \frac{m\,d\dot{\varepsilon}}{\dot{\varepsilon}\,d\varepsilon} - 1 = 0$$

It gives:

$$= \frac{n}{\varepsilon} + m - 1 = 0$$

Instability during necking:

$$\varepsilon = \frac{n}{1-m} \tag{4.14}$$

Rational strain to instability with work hardening of value n and coefficient m of sensitivity of the stress to the strain rate (for constant imposed tensile speed).

NOTE.– We found that n and m can be obtained by Hopkinson-type dynamic tests (see Chapter 5, section 5.4).

When the influence of the true strain rate $\dot{\varepsilon}$ on the stress $\sigma(\dot{\varepsilon})$ is negligible, the condition of instability $\sigma \alpha A + A \alpha \sigma = 0$ implies a threshold for necking to appear $\varepsilon_S = n$. On the other hand, when the influence of the speed becomes appreciable, and if m represents the change of σ as a function of $\dot{\varepsilon}$ (m = dLogσ/dLog$\dot{\varepsilon}$), the necking works to cause a strain $\varepsilon_S = n/(1 - m)$.

4.1.5. Summary

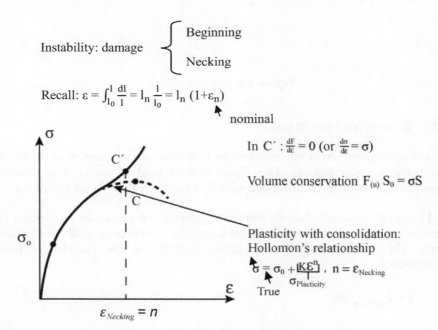

Figure 4.4. *Stress and strain*

$$\sigma_{necking} = Fe^{\varepsilon \, necking}$$

$$\frac{F}{Fnecking} = \left(\frac{\varepsilon}{n}\right)^n e^{n-\varepsilon}$$

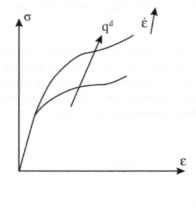

$\sigma_p = K' \varepsilon^n \dot{\varepsilon}^m$

With m: coefficient of sensitivity of the stress at the speed

or in force: $F = K' \varepsilon^n \dot{\varepsilon} \ e^{-\varepsilon}$

If the elongation speed of the test sample through a tensile test is imposed: $\frac{\Delta l}{dt} = \text{Const.}$

we obtain: $\dot{\varepsilon} = \dot{\varepsilon}_n \ e^{-\varepsilon}$

where: $\frac{d\dot{\varepsilon}}{d\varepsilon} = \dot{\varepsilon}$

at instability $\frac{dF}{d\varepsilon} = 0$

$\varepsilon = \frac{n}{1-m}$

Figure 4.5. *Effect of speed*

4.1.6. *Generalized strain speed*

The generalized strain rate provides an indication of the intensity of the strain rate using a single number. This figure will be used whenever it is deemed appropriate to simulate a forming operation by a mechanical test in the laboratory.

Finally, if we accumulate the elementary generalized strains $\dot{\bar{\varepsilon}}dt$ an element of particles undergoes over time, we define a quantity $\bar{\varepsilon}$, referred to as the generalized strain. The integral must be calculated by following the particles in their movements:

$$\dot{\bar{\varepsilon}} = \int_{\text{particle}} \dot{\bar{\varepsilon}}dt$$

The generalized strain $\bar{\varepsilon}$ will be used to characterize by a single number the "amount of strain" an element of matter undergoes.

With a single number, we define the intensity of the strain rate. This is the generalized strain speed $\dot{\bar{\varepsilon}}$ (pronounced "epsilon barre point").

For primary directions:

$$\dot{\bar{\varepsilon}} = \sqrt{\frac{2}{3} \left(\dot{\varepsilon}_I^2 + \dot{\varepsilon}_{II}^2 + \dot{\varepsilon}_{III}^2 \right)}$$

The sum in parentheses is equal to the sum of the diagonal terms of the square of the matrix in the principal directions. This sum is a tensor invariant. Thus, in any direction:

$$\dot{\bar{\varepsilon}} = \sqrt{\frac{2}{3} \left(\dot{\varepsilon}_{xx}^2 + \dot{\varepsilon}_{yy}^2 + \dot{\varepsilon}_{zz}^2 + 2\dot{\varepsilon}_{xy}^2 + 2\dot{\varepsilon}_{yx}^2 + 2\dot{\varepsilon}_{zx}^2 \right)} = \sqrt{\frac{2}{3} \sum_i \sum_j \dot{\varepsilon}_{ji} \cdot \dot{\varepsilon}_{ij}}$$

If we wish to characterize the amount of strain an element undergoes, we calculate the *generalized strain* (epsilon bar):

$$\bar{\varepsilon} = \sqrt{\frac{2}{3} \sum_i \sum_j \varepsilon_{ji} \varepsilon_{ij}}$$

4.2. Forming by pressing

4.2.1. *Study of plastic instability; influence of work hardening and anisotropy (characteristics of pressing thin sheets[1])*

4.2.1.1. *Principle, instability on the basis of n and r*

EXAMPLE 4.1.–

A blank is held in place between a circular die and a blank clamp (Figure 4.6). A ring can be used to prevent the sheet from sliding radially. A fluid under a pressure of p swells the blank. As the swelling progresses, the curve increases; this test removes the punch, thus eliminating friction between the punch and the sheet.

Recall:

$$\varrho = \frac{a^2 + h^2}{2h}, \ \sigma_r = \sigma_I = \sigma_{II} = \frac{P\varrho}{2e} \text{ and } \sigma_{III} = 0$$

1 Parnière, P., Sanz, G. Édition du CNRS.

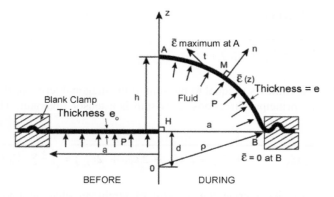

Figure 4.6. *Diagram of a biaxial expansion test (Jovignot)*

4.2.1.2. *Application of the Von Mises criterion*

As the thickness of the sheet e is very small in front of the radius of curvature ϱ, we can negate the normal stress σ_n and assume that a plane state of stress exists. The Von Misis criterion can then be given as:

$$J_2 = k_{VM}^2 = \frac{1}{6}[(\sigma_I - \sigma_{II})^2 + (\sigma_{II} - \sigma_{III})^2 + (\sigma_{III} - \sigma_I)^2]$$

$$J_2 = \frac{1}{6}[2\,\sigma_I^2] = \frac{1}{3} \quad 0 \quad \frac{1}{3}\sigma_r^2$$

In tension: $\sigma_I = \sigma_0$, $\sigma_{II} = \sigma_{III} = 0$.

$$J_2 = \sigma_0^2/3$$

where $\sigma_r = \sigma_0$ (VM).

At all points:

$$P = \frac{2\sigma_0 e}{\varrho} \tag{4.15}$$

Instability will occur when the generalized strain $\bar{\varepsilon}$ is such that:

$$\frac{dp}{p} = 0 \tag{4.16}$$

4.2.1.3. *Calculation of* $\bar{\varepsilon}$

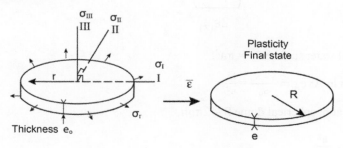

Figure 4.7. *Circumferential stretching*

$$\varepsilon_I = \varepsilon_{II} \qquad\qquad \text{Plasticity} \qquad\qquad e_0 \text{ becomes } e$$
$$\xrightarrow{} \qquad\qquad \varepsilon_I + \varepsilon_{II} + \varepsilon_{III} = 0$$
$$\text{diV } \vec{D} = 0 \qquad\qquad\quad |\quad\quad |$$
$$\qquad\qquad\qquad\qquad\qquad\qquad \varepsilon \quad\quad \varepsilon$$

$$\varepsilon_{III} = -\,(\varepsilon_I + \varepsilon_{II}) = -2\varepsilon$$

$$\bar{\varepsilon} = \frac{\sqrt{2}}{3}\sqrt{(\varepsilon_I - \varepsilon_{II})^2 + (\varepsilon_{II} - \varepsilon_{III})^2 + (\varepsilon_{III} - \varepsilon_I)^2} = \frac{\sqrt{2}}{3}\sqrt{18\varepsilon^2} = 2\varepsilon = -\,\varepsilon_{III} \qquad [4.17]$$

$$\Phi_1^2 \qquad\qquad\quad \Phi_2^2 \qquad\qquad\quad \Phi_3^2$$
$$\downarrow \qquad\qquad\qquad \downarrow \qquad\qquad\qquad \downarrow$$
$$0 \qquad\qquad\quad (3\varepsilon)^2 \qquad\qquad (-3\varepsilon)^2$$

4.2.1.4. *Calculation of* $\bar{\varepsilon}_A$ *and* $\bar{\varepsilon}_Z$ *(in M)*

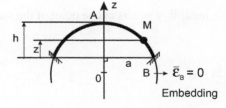

Figure 4.8. *Spherical dome*

The pellet undergoes a state of stresses within its plane of the value $\sigma_r = \dfrac{P\,\varrho}{2e}$, and thus a biaxial expansion in its plane (I, II) with, in M(z):

$$\varepsilon_I = \varepsilon_{II} = \varepsilon_{(z)}$$

According to the thickness $\varepsilon_{III} = -\underbrace{(\varepsilon_I - \varepsilon_{II})}_{2\varepsilon_{(z)}}$:

– Initial state: surface $S_0 = \pi a^2$;

– strain in A: $e_A = \dfrac{S - S_0}{S_0} = \dfrac{h^2}{a^2}$;

– state at h: spherical dome $S = 2\pi\varrho h$.

$$e_A = \frac{h^2}{a^2}$$

And according to z (proportional strain):

$$e_z = \frac{hz}{a^2} \text{ with } 0 \le z \le h \; z = 0 \to \varepsilon_B = 0$$

$$z = h \; \to \; \varepsilon_A = \frac{h^2}{a^2}$$

For the "points" M located in z, we obtain:

$$\bar{\varepsilon}_{(z)} = 2\varepsilon_z = 2L_n\left(1 + \frac{hz}{a^2}\right) = 2\varepsilon_{I_{(z)}} = 2\varepsilon_{II_{(z)}} \qquad [4.18]$$

And:

$$\varepsilon_{III_{(z)}} = -\bar{\varepsilon}_{(z)}$$

4.2.1.5. *Instability*

With [4.15]: $P = \dfrac{2\sigma_0 e}{\varrho}$, instability occurs at one point of the stamped material with [4.16]: $\dfrac{dP}{P} = 0$, or:

$$\frac{dP}{P} = \frac{d\sigma_0}{\sigma_0} + \frac{de}{e} - \frac{d\varrho}{\varrho} = 0 \qquad [4.19]$$

For a rigid plastic body $\dfrac{d\sigma_0}{\sigma_0} = 0$.

With the strains reaching maximums at the pole A or z = h.

$$\bar{\varepsilon}_A = 2\,\varepsilon_{z=h} = -\varepsilon_{III,z=h} \qquad [4.20]$$

thus:

$$\frac{de}{e} = d(2\varepsilon)_z = -d\overline{\varepsilon}$$ [4.21]

and:

$$\frac{d\varrho}{\varrho} = \frac{d\overline{\varepsilon}}{2} - \frac{1}{4}\frac{e^{\overline{\varepsilon}/2}d\overline{\varepsilon}}{e^{\overline{\varepsilon}/2}-1}$$ [4.22]

thus, the condition of instability with [4.19]:

$$\frac{de}{e} = \frac{d\varrho}{\varrho}$$ [4.23]

$$\frac{3}{2} - \frac{1}{4}\frac{e^{\overline{\varepsilon}/2}}{e^{\overline{\varepsilon}/2}-1} = 0$$ [4.24]

and:

$$\overline{\varepsilon} \simeq \frac{4}{11} = 0.36 \text{ in A}$$ [4.25]

4.2.2. *Parameters influencing instability: influences of the work hardening n and the anisotropy r*

4.2.2.1. *The work hardening coefficient[2] n*

The shape of the plastic part of the rational tensile curve is described fairly accurately (extra-mild steel, ferritic steel, in general) by the law governing the shape:

$$\sigma = K\,\varepsilon^n$$ [4.26]

For alloy steels, austenite steels, light alloys, etc., it is necessary to consider more complex laws (Pomey), in the form of, for example:

$$\sigma = A + K\,\overline{\varepsilon}^n$$ [4.27]

4.2.2.2. *The anisotropy coefficient[3] r*

This coefficient is used as a characteristic of plastic anisotropy in Hill's theory. It depends on the crystallographic texture of the sheet metal.

---

2 See Table 4.1.
3 See Table 4.2.

Under tension, the anisotropy coefficient is equal to the ratio of the rational strains in the two directions orthogonal to the direction of tension:

$$r\left(\varepsilon_1, \alpha\right) = \frac{\varepsilon_2}{\varepsilon_3}$$

For an isotropic material, we have $\varepsilon_2 = \varepsilon_3 = -\varepsilon_1/2$ giving:

$$r\left(\varepsilon_1, \alpha\right) = 1$$

For a non-isotropic material $r\left(\varepsilon_1, \alpha\right) \neq 1$.

Most metals have a coefficient r, which depends on their elongation. An average anisotropy coefficient is defined $\bar{r}$ from the plane of the sheet metal by:

$$\bar{r} = \frac{1}{2\pi} \int_0^{2\pi} r(\alpha)\, d\alpha$$

For a work hardening law $\sigma_0 = K\, \bar{\varepsilon}^n$, we obtain:

$$\frac{d\sigma_0}{\sigma_0} = n \frac{d\bar{\varepsilon}^n}{\bar{\varepsilon}^n} \qquad\qquad [4.28]$$

Metal	n
Extra-mild steel	0.15 to 0.25
Ferritic steel at 17% Cr	0.16 to 0.20
Austenitic steel 18-10	0.40 to 0.50
Aluminum	0.07 to 0.27
AG 1 to 5	0.23 to 0.30
A S G	0.23
AU$_4$ G	0.15
Brass 67 Cu – 33 Zn	0.55
Brass 63 Cu – 37 Zn	0.45
Copper	0.30 to 0.47
Zinc	0.1
Nickel	0.6

Table 4.1. *Values of n for different materials*

Metal	$\bar{r}$
Extra-mild steel	1 to 2
Ferritic steel at 17% Cr	0.9 to 1.3
Austenitic steel 18-10	0.9 to 1
Aluminum	0.5 to 1
AG 1 to 5	0.6 to 0.8
A S G	0.6
AU_4 G	0.7
Brass 67 Cu – 33 Zn	1
Brass 63 Cu – 37 Zn	0.85
Copper	0.9 to 1
Zinc	0.5
Nickel	1

Table 4.2. *Values of $\bar{r}$ for different materials (Pomey and Grumbach)*

[4.19] condition of instability with [4.24]:

$$\frac{n}{\bar{\varepsilon}} = \frac{3}{2} - \frac{1}{4} \frac{e^{\bar{\varepsilon}/2}}{e^{\bar{\varepsilon}/2}-1}$$

[4.29]

That is:

$$\bar{\varepsilon} \simeq \frac{4}{11}(2n + 1)$$

[4.30]

The increase in the work hardening coefficient favors biaxial expansion, since it delays the occurrence of instability. This result is consistent with a great deal of experimental observations, but plastic anisotropy does have influence, and the increase in $\bar{r}$ decreases ε_{III}, thus from $\bar{\varepsilon}$ [4.20]. On the other hand, the height h of the stamped section has little variance with $\bar{r}$ since an increase in this variable tends toward greater homogeneity in the strains in the thickness (see Figure 4.9).

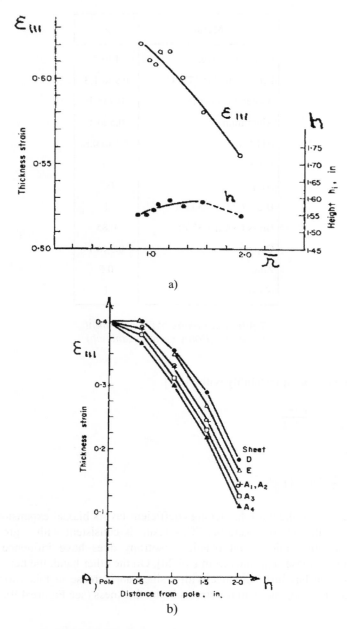

Figure 4.9. *Changes in thickness and depth on the basis of n and $\bar{r}$ (according to Horta et al. 1970)*

COMMENT ON FIGURE 4.9.– *(a) Influence of $\bar{r}$ on the height of a Jovignot essay and the strain in thickness to instability (extra-mild steel). b) Radial distribution of strains in thickness ($\varepsilon_{n\,pôle} = 0.4$) as a function of $\bar{r}$, sheet metal D: $\bar{r} = 1.95$; sheet metal A_2: $\bar{r} = 1.20$; sheet metal E: $\bar{r} = 1.55$; sheet metal A_3 : $\bar{r} = 0.90$; sheet metal A_1: $\bar{r} = 1.35$; sheet metal A_4: $\bar{r} = 1.35$.*

4.2.3. *Pressing and formability in bending of sheets, shaping of elliptical bulbs*

EXAMPLE 4.2.–

The bulb of rays R_1 and R_2 undergoes a biaxiality in stress σ_I, σ_{II} ($\sigma_{III} = 0$) under the action of the pressure P of a fluid (Figure 4.10); we obtain:

$$\bar{\sigma} = (\sigma_I^2 + \sigma_{II}^2 - \sigma_I\sigma_{II})$$

In plasticity, we obtain:

$$\varepsilon_I + \varepsilon_{II} + \varepsilon_{III} = 0$$

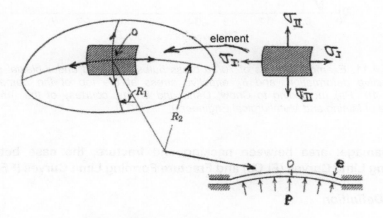

Figure 4.10. *Forming of bulb by pressing*

And:

$$\bar{\varepsilon} = \sqrt{2}/3 \, [(\varepsilon_I - \varepsilon_{II})^2 + (\varepsilon_{II} - \varepsilon_{III})^2 + (\varepsilon_{III} - \varepsilon_I)^2]^{1/2}$$

At the equilibrium of the sheet metal element:

$$P/e = \sigma_1/R_1 + \sigma_{II}/R_2$$

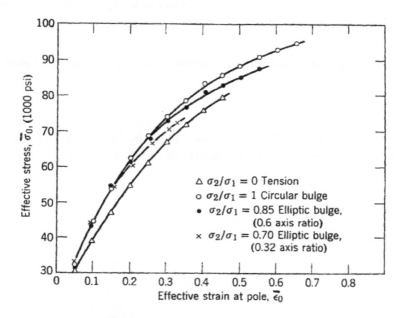

Figure 4.11. *Example of tests of 70/30 brass bulbs annealed under biaxial stress, $\bar{\varepsilon}_0$ effective deformation, and $\bar{\sigma}_0$ effective stress at the top of the dome (Psi: 6.89×10^3 Pa) (according to Chow, Dana and Sachs, courtesy of the American Institute of Mining and Metallurgical Engineers)*

4.3. Damage: area between necking and fracture, the case between Forming Limit Curves (FLCs) and Fracture Forming Limit Curves (FFLCs)

4.3.1. *Definition*

$$S_D = S - \tilde{S} \rightarrow \text{resistant surface}$$

S of defects   S total

$\frac{S_D}{S} = D$ is the damage.

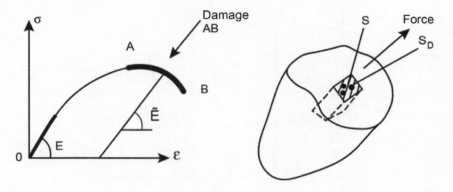

Figure 4.12. *Damage*

– Stress: force $F = \sigma S = \tilde{\sigma} \tilde{S}$:

$$\tilde{\sigma} = \sigma \frac{S}{\tilde{S}} = \sigma \frac{S}{S - S_D} = \frac{\sigma}{1 - D}$$

$\tilde{\sigma} = \frac{\sigma}{1 - D}$ (effective stress on a damaged material)

– Strain:

$$\varepsilon_{damaged} = \frac{\tilde{\sigma}}{E} \rightarrow \varepsilon_{damaged} = \frac{\sigma}{(1-D)E}$$

with $\tilde{E} = (1\text{-}D)$ E modulus or stiffness of the damaged material, we have $\varepsilon_{damaged} = \frac{\sigma}{\tilde{E}}$ and $\tilde{E} = (1\text{-}D)$ E gives $\tilde{E} = E - D.E$.

We have the damage as: $D = \frac{E - \tilde{E}}{E} = 1 - \frac{\tilde{E}}{E}$.

4.3.2. *Damage measurement D*

– Mechanical test:

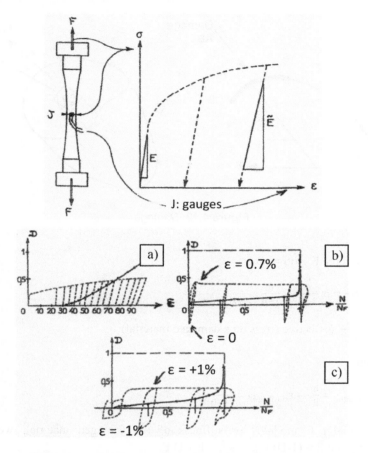

Figure 4.13. *Examples of the change in damage (Benounich 1990)*

COMMENT ON FIGURE 4.13.– *(a) Damage to ductile plastic, copper, T = 20°C, $\dot{\varepsilon} \sim 10^{-4} S^{-1}$. (b) Fatigue damage in cyclic test, steel 316 L, T = 20°C:*

$$\varepsilon = \Bigg\langle \begin{array}{l} + 0.7 \times 10^{-2} \\[6pt] \qquad N_F \ (fracture) = 70,450 \ cycles \\[6pt] 0 \end{array}$$

c) Damage of cyclic creep 316L steel, T = 550°C:

$$\varepsilon = \Bigg\langle \begin{array}{l} + 10^{-2} \\[6pt] \qquad N_c (fracture) = 218 \ cycles \\[6pt] -10^{-2} \end{array}$$

Ultrasonic wave for measurement of damage:

– Wave speeds:

$$V_L^2 = \frac{E}{\varrho} \frac{1-v}{(1+v)(1-2v)}$$

– Damaged:

$$\widetilde{V}_L^2 = \frac{\widetilde{E}}{\varrho} \frac{1-v}{(1+v)(1-2v)}$$

with:

$$\widetilde{E} = \varrho \, \widetilde{V}_T^2 \, \frac{3\widetilde{V}_L^2 - 4\widetilde{V}_T^2}{\widetilde{V}_L^2 - V_T^2}$$

$$D = 1 - \frac{\widetilde{E}}{E} = 1 - \frac{\widetilde{V}_L^2}{V_L^2}$$

EXAMPLE 4.3.–

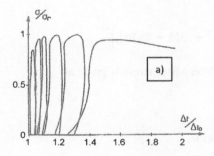

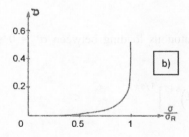

Figure 4.14. *Damage measurement of concrete by ultrasound (Benounich 1990). (a) Change in the travel time of the wave $\frac{\Delta t}{\Delta t_0} = \frac{V_L}{\widetilde{V}_L}$ as a function of the stress related to the fracture stress. (b) Change of the damage*

4.3.3. *Large strains and damage*

Damage:

– cracks, cavities, etc. → decrease in stiffness (E varies);

– E: modulus of elasticity of the virgin material;

– $\widetilde{E}$: elasticity modulus of the damaged material.

We define the damage D by:

$\widetilde{E} = E\,(1\text{-}D)$

That is:

$$D = 1 - \frac{\widetilde{E}}{E} \qquad\qquad [4.31]$$

During successive loading and unloading tests, the evolution of $\widetilde{E}$. We will refer to σ^* as an equivalent stress in the sense of the damage, and defined by the following relation:

$$\sigma^* = \left[\frac{2}{3}\,(1 + \nu)\bar{\sigma}^2 + 3(1 - 2\nu)\sigma_m^2\right]^{1/2}$$

with $\bar{\sigma}$, the equivalent Von Mises stress is given as:

$$\bar{\sigma} = (\tfrac{2}{3}\,S : S)^{1/2}$$

with S, the stress diverter σ:

$$\sigma_m : \frac{1}{3}\,\text{traces}\,(\sigma)$$

For an increasing monotonous loading between $\sigma^* = 0$ and σ^*maxi, at every point:

$$D = 1 - \left[1 - \left(\frac{\sigma^* - \sigma_d}{S}\right)^{s+1}\right]^{1/s+1}$$

with:

– σ_d: damage threshold;

– S, s: characteristic constants to be identified with a tensile test.

4.3.4. *Stress curve, strain of a 30 CD4 steel that has undergone a perlite globularization annealing (as given by Gathouffi (1984))*

EXAMPLE 4.4.–

$$\sigma = \sigma_y + K\,(\bar{\varepsilon}_p)^n$$

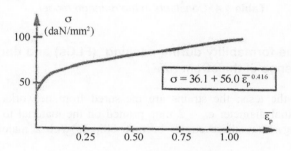

Figure 4.15. *Tensile test on 30 CD4*

Initial Young's modulus E_0	246,600 Mpa
Poisson's Ratio ν	0.264
Limit σ_y	361 MPA
K	560 MPa
n	0.416

Table 4.3. *Constants of the material*

4.3.4.1. *Test with successive loading and unloading: damage*

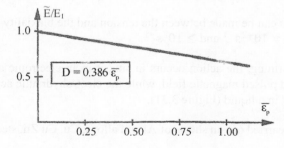

Figure 4.16. *Evolution of the Young's modulus*

Damage threshold σ_d	361 MPa
S	701 MPa
s	1.042
δ	0.386

Table 4.4. *Constants in the damage model*

4.4. Limit of the formability during necking (FLCs) and during fracture (FFLCs): influence of the strain rate

Principle of the tests: the strains are measured from networks of interlaced circles, of an initial diameter $e_0 = 2$ mm, printed on the material to be deformed. After undergoing strain, these circles become ellipses with a major axis e_1 and minor axis e_2.

The formability limit curves are shown in the reference $(\varepsilon_1, \varepsilon_2)$ of rational extensions:

$$\varepsilon_1 = \text{Log}\frac{e_1}{e_0}, \varepsilon_2 = \text{Log}\frac{e_2}{e_0}$$

The right part of the curves (in the first quadrant of the reference) $(\varepsilon_1, \varepsilon_2)$ gives the aptitude for biaxial-type strain under stress, the central domain (axis of ε_1) indicates the behavior of the metal in plane strain and the left part (ε_2 negative) represents the tensile formability (Figure 4.17).

The plates are stamped by a press punch or a fluid (static strain), by an intense and pulsed magnetic field or by electrohydraulic effect (strain by shock). These last two techniques make it possible to strain metals at high speeds of strain $\dot{\varepsilon} > 10^2 s^{-1}$, and the pressing speed can be in the order of 200 m/s.

A comparison can be made between the tension and the biaxiality in necking and during fracture $\dot{\varepsilon} \simeq 10^{-2} s^{-1}$ and $\geq 10^2 s^{-1}$.

In magnetoforming, the action occurs in situ, it is an electronic action obtained by an intense and pulsed magnetic field, while the electrohydraulic action is done by "lightning effect" in a liquid (Figure 3.31).

The tests are carried out on sheets of Al and alloys, Cu, Cu Zn, steel, etc.

The FLCs allow for the success of the shaping to be predicted, particularly the influence on the strains obtained for the lubricants, the characteristics of the metal sheets and the operating conditions.

4.4.1. *Types of strain*

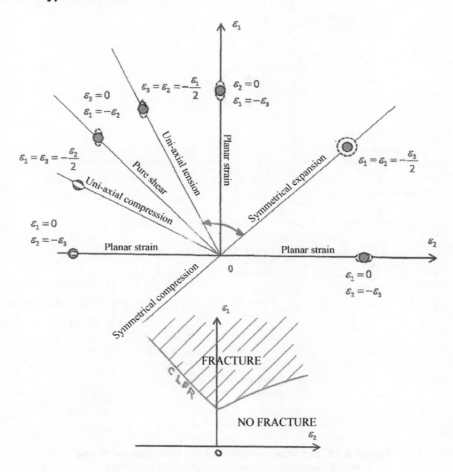

Figure 4.17. *Types of flat stresses and fracturing formability limit curve*

We know how to measure, track and modify the strains of a piece during its formation, but we lack the knowledge of the limits of strains of the metal for the most common direct and complex trajectories in pressing.

This necessity requires us to create laboratory tests to simulate these different trajectories and determine the limit curves of the formation of the primary commercial qualities (curves: Forming Limit Curve (FLC)).

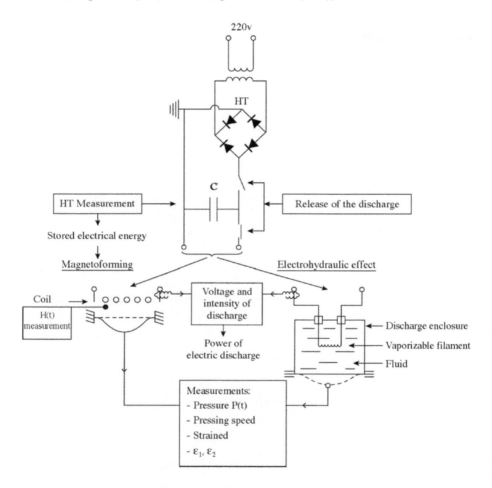

Figure 4.18. *Schematic diagram of the experimental device which allows the formability limit of the materials at high strain speeds to be obtained*

NOTE.– The influences of the work hardening coefficient n and the anisotrophic coefficient r are as follows:

– the increase in the work hardening coefficient n reflects an increase in ductility with an increase in the FLCs and the FFLCs;

– the anisotropy coefficient r has a marked influence on the trajectories.

$\varepsilon_1 = f(\varepsilon_2)$, thus an interaction with the FLCs is dependent on R.

The slope α from the beginning of the trajectories is accurately given by the relation:

$$\alpha = -\frac{r+1}{r}$$

4.4.1.1. Test sample for FLCs

EXAMPLE 4.5.–

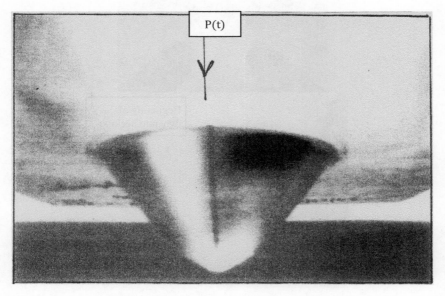

Figure 4.19A. *Free stamped by magnetic field. Photos: The required formability curves (FLCs and FFLCs, necking fracture) in the dynamic pressing of etched sheets stressed by electromagnetic induction shocks and/or by the electrohydraulic effect (Leroy)*

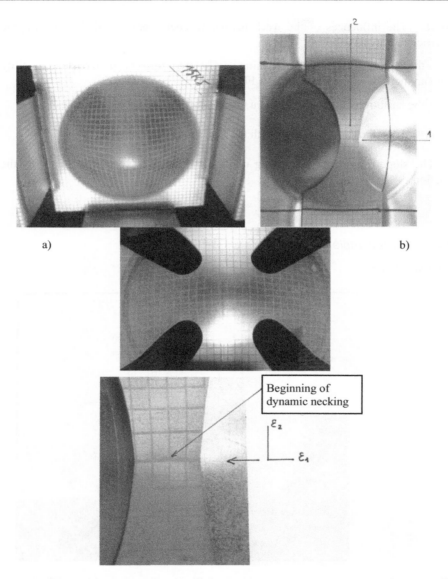

Figure 4.19B. *(a) Biaxial and planar expansions. (b) Dynamic tension. Photos: The formability limit curves (FLCs and FFLCs, necking fracture) in the dynamic pressing of etched sheets stressed by electromagnetic induction shocks and/or by the electrohydraulic effect (Leroy)*

4.4.2. Comparison of FLC by $\dot{\varepsilon}$: shock by electronic action in the case of aluminum alloys

EXAMPLE 4.6.–

Figures 4.20–4.22 are taken from Renaud and Leroy (1975) (article[4] by Renaud and Leroy, transmitted by Néel).

Viscoplastic behavior of polycrystalline aluminum alloys under high pulsed magnetic field (formability limit curves).

Metalforming process by high pulsed magnetic field has been used for the investigation of polycrystals viscoplastic behavior at high strain rate ($\dot{\varepsilon} \simeq 10^2 s^{-1}$). The formability limit curves concerning two aluminum alloys with the appearance of localized necking and fractures are compared to those obtained at low strain rate.

The fact that a number of works carried out on the high-speed strain ($\dot{\varepsilon} \simeq 10^2 s^{-1}$) of aluminum single crystals by magnetic fields (inductions in the order of 10 Teslas) have shown notable differences in the stress–strain curves compared to those obtained at conventional speeds has led us to research the strain limit curves of fine-grained polycrystals ($d \simeq 10 \mu$, anisotropy coefficient $r \simeq 1$) and to carry out a comparison with those obtained during slow strains. The comparative curves are given for two alloys of aluminum: A (5052 H 34) and B (9004 H 36), which have the following compositions by weight:

Figure 4.20. *Comparison of formability limits. Influence of stress speeds (Renaud and Leroy 1975)*

	Cu	Mg	Mn	Cr	Fe	Si
A	0.02	2.30	0.04	0.24	0.32	0.15
B	0.14	0.89	0.98	0.02	0.55	0.28

Since the work of Goodwin and Keeler, the strain limit curves have been used to characterize the flow suitability of thin plates. They indicate the maximum strain that a metal can undergo when it is subjected to various types of stresses, ranging from pure tension to biaxial expansion. The boundary curves at the fracture and those at the localized appearance of the necking are traced for the references ε_1, ε_2 (with ε_1 and ε_2 being the rational elongations measured in the main directions of the strain). Since the different techniques that make it possible to obtain the limit of the necking give essentially identical results, we have chosen the Bragard method.

---

4 Session of June 9, 1975.

To scan a wide area of the plane ε1, ε2, we strained thin flat test samples (thickness of 0.25 mm, lower thickness than the studied alloys) of different widths in a circular matrix (ø = 80 mm) in order to obtain results comparable to those given by the Nakazima method.

The strains are measured from a network of 2 mm circles printed onto the metal using a photographic process. We provide the figures for the limit curves obtained during the strain by magnetic fields (Figures 4.27A and B) and we compare them to the curves obtained during slow strains.

From these curves, we can make the following remarks:

– The shape of the boundary curves in the necking is very close to that of the theoretical curves, especially in the domain of 2 < 0. These curves are straight lines with a slope close to -1.

– The flow limit curves move back toward an improvement in their suitability for strain, both for fracture and necking (with an improvement in the order of 50% for polycrystal A and 35% for polycrystal B).

– The domain that extends between the appearance of the localized necking and the fracture is greatly increased (63% for A and 31% for B).

Figure 4.21. *Comparison of formability limits. Influence of stress speeds (Renaud and Leroy 1975)*

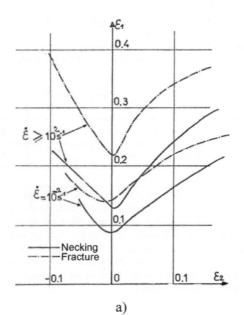

a)

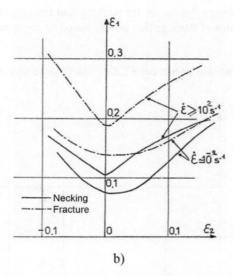

b)

Figure 4.22. *(A) Influence of the strain rate for the case of the polycrystal A (Al 5052 H34) (as given by Renaud and Leroy 1975). (B) Influence of the strain rate for the case of the polycrystal B (Al 3004 H36) (as given by Renaud and Leroy 1975)*

ADDITIONAL MATERIALS.– Chiem, C.Y., Renaud, J.-Y., Leroy, M. (1975). Compte rendu. *Acad. Sc.*, 280, Series B.

Leroy, M. (1972). Thesis, University of Nantes.

Martin, R. (1972). *Cétim Informations*, December 28, 49–61.

Vigier, P. and Lefort, S. (1975). Compte rendu. Contrat D.G.R.S.T., no. 73-7-1593.

The experimental curves generally show an increase in the ability to be formed at high strain rates, varying according to the materials and the types of strains between uniaxial tension by tension:

$$\varepsilon_3 = \varepsilon_2 = -\varepsilon_1/2$$

planar strain $\varepsilon_2 = 0$, and:

$$\varepsilon_1 = -\varepsilon_3$$

and the symmetric expansion, giving:

$$\varepsilon_1 = \varepsilon_2 = -\varepsilon_3/2$$

The separation between the curves for necking and fracture in static and dynamic forms is a good indicator of damage that occurs based on the types of strain.

4.4.3. *Influence of strain rates on FLCs: static and dynamic formability*[5]

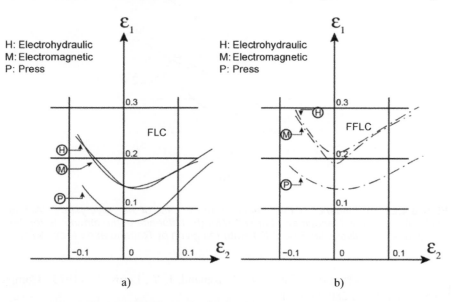

Figure 4.23. *Al 3004 (AM1G), H36 (3/4 hard); in (a) FLC (necking);
in (b) FFLC (fracture); ep = 0.3 mm (Renaud and Leroy 1975)*

NOTE.– For industrial applications, it is necessary to avoid damage, and therefore carry out stress trajectories in the domain $(\varepsilon_1, \varepsilon_2)$, which is lower than the FLC curves.

Plastic instability is very important in processes such as the pressing of metal sheets. It is clearly necessary to choose both materials as well as shaping techniques and tools in such a way as to carry out strain trajectories without causing damage.

Mild steel is a good material for pressing because it undergoes a high level of plastic strain before necking: thus, we can press it very deeply without breaking it.

Aluminum alloys are much less advantageous, and as a result, they can only be pressed a moderate amount before necking. Pure aluminum is somewhat less

---

5 Renaud and Leroy (1975).

unfavorable, but it is much too soft for most applications. The high strain speeds (dynamic plasticity) help to increase the flows before necking.

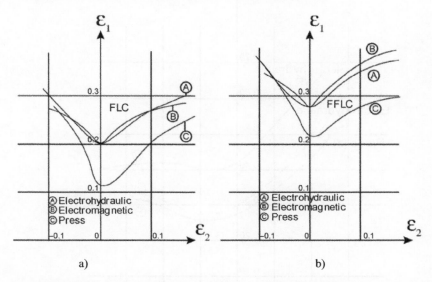

a)   b)

Figure 4.24. *Work-hardened copper, ep = 0.3 mm; in (a) FLCs; in (b) FFLC (Renaud and Leroy 1975). (a) Limit curves of necking FLCs. (b) Limit curves of fracture FFLC by press P (low speed) and high speed tests ($\dot{\varepsilon} \geq 10^2 s^{-1}$)*

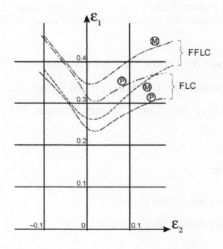

Figure 4.25. *Work-hardened brass Cu Zn 36, thickness = 0.3 mm (Renaud and Leroy 1975)*

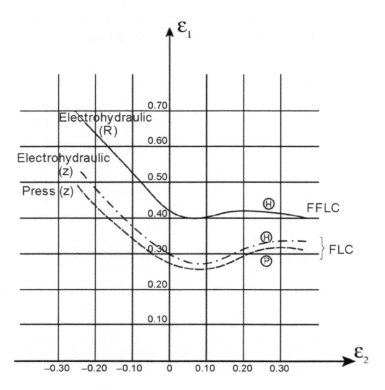

Figure 4.26. *Steel-type XES pressing,
thickness = 0.6 mm (Renaud and Leroy 1975)*

Following the low and high speed tests, we give the results for:

– the change in the trajectories $\varepsilon_1 = f(\varepsilon_2)$ toward the necking Z and the fracture R;

– the changes in $Z(\varepsilon_2)$, $R(\varepsilon_2)$ and the gap between FLC and FFLC domains involving the damage.

4.4.4. *Metals studied*

The following metals have been studied:

– annealed aluminum A99, with thickness 1 mm;

– Ag 2.2 (5052) and AMIG (3004) alloys, with thickness 0.3 mm;

– deoxidized work-hardened copper Cu/b, with thickness 0.3 mm;

– work-hardened CuZn 36 brass, with thickness 0.3 mm;

– pressing steel of type XES, with thickness 0.6 mm.

4.4.5. *Strain trajectories: change of ε_1 under ε_2; comparison of low and high speeds*

4.4.5.1. *The case of Al Ag 2.2 (5052) and Cu*

EXAMPLE 4.7.–

Upon determining the influence of the strain trajectories on the formability limit curves, we have plotted the trajectories at several points of the limit curves for the aluminum alloy Ag 2.2 (5052) strained by a magnetic field and for copper strained by electrohydraulic discharges. These trajectories are compared with those obtained at low strain speeds in the figures.

NOTE.– It should be noted that in high-speed strain, each point of the trajectory is not obtained with the same strain speed, since this depends on the energy involved to obtain the desired strain rate. In the figures, we note the average strain speeds $\bar{\dot{\varepsilon}}$ at which each point of the trajectories has been reached. However, all these speeds fall between 100 s^{-1} and 300 s^{-1}, and it does not apppear that the trajectories are significantly influenced by these variations.

4.4.5.2. *Results obtained on these trajectories*

A comparison of the trajectories at high- and low-strain speeds does not show any noticeable differences. In both cases, the points of the boundary curve are obtained with linear trajectories. Up to the point of necking (z), the trajectories are straight lines passing through the origin, then from the necking to the fracture (R), vertical straight lines (Figure 4.27):

– from 0 to Z: $\varepsilon_1 \simeq k\varepsilon_2$;

– Z to R: $\varepsilon_2 \simeq$ const.

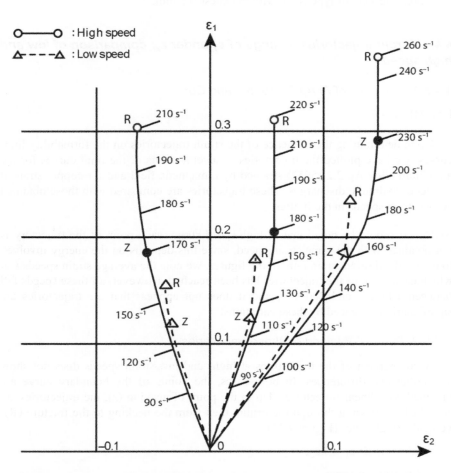

Figure 4.27. *(A) Comparison of the trajectories of strains at low and high speeds by magnetoforming for alloy 5052 and electrohydraulic shock for copper. Z: necking; R: fracture (Renaud and Leroy 1975)*

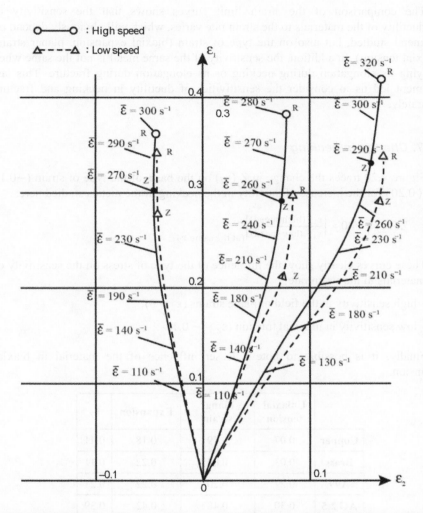

Figure 4.27. *(B) Comparison of the trajectories of strains at low and high speeds, by magnetoforming for alloy 5052, by electrohydraulic shock for copper. Z: necking; R: fracture (Renaud and Leroy 1975) (continuation)*

4.4.6. *Comparison of the values for necking Z and fracture R*

The comparison of the strain limit curves shows that the sensitivity of the ductility of the materials to the strain rate varies, which will obviously depend on the metal studied, but also on the type of strain (biaxial expansion, planar strain, uniaxial tension). In addition, the sensitivity of the same metal is not the same when studying its elongation during necking or its elongation during fracture. This last comment led us to consider the sensitivities of ductility in necking and fracture separately.

4.4.7. *Change in necking*

Figure 4.28 traces the change in Z (ε_2) on the basis of the type of strain (-0.10 ⟨ ε_2 ⟨ 0.20) for the increase of ductility in the necking of the metals studied here:

$$Z\,(\varepsilon_2) = 100 \times \left[\frac{\varepsilon_{1Z}\ \text{(high speed)}}{\varepsilon_{1Z}\ \text{(low speed)}}\right]_{\text{(at the same } \varepsilon_2)}$$

These curves clearly show the influence of the type of stress on the sensitivity of the materials to the strain rate:

– high sensitivity in the field of plane strains ($\varepsilon_2 = 0$);

– low sensitivity in uniaxial traction ($\varepsilon_2 = -0.1$).

Finally, it is possible to note a lesser influence of the material in biaxial expansion.

	Uniaxial tension	Planar strain	Expansion	$\bar{a}_Z$
Copper	0.07	0.79	0.18	0.41
Brass	0.03	0.08	0.22	0.11
A99	0.27	0.20	0.22	0.21
AG 2.5	0.30	0.48	0.42	0.39
AM1G	0.34	0.71	0.20	0.44
Steel	0.07	0.07	0.08	0.07

Table 4.5. *Comparison of necking (Leroy)*

In order to better understand these various influences, we have defined a coefficient a_z that describes the sensitivity of the ductility in necking at the rate of strain for the following metals:

$$a_z = \left[\frac{\varepsilon_{1Z} \text{ (high speed)}}{\varepsilon_{1Z} \text{ (low speed)}} - 1 \right]_{\varepsilon_2 = c^{te}}$$

and an average coefficient $\bar{a}_Z$:

$$\bar{a}_Z = \frac{1}{n} \sum_{(\varepsilon_2)}^{n} . a_z$$

The values of these coefficients are given in Figure 4.28:

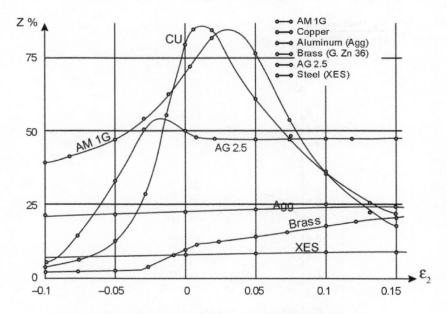

Figure 4.28. *Change of ductility in necking. Sensitivity of materials to the strain rate and according to the type of stress $(- 0.1 \langle \varepsilon_2 \langle 0,2)$ (Leroy)*

It appears that the sensitivity coefficient in necking is even weaker since the metal has good pressing characteristics under pressure (this is the case of for brass, annealed aluminum and XES steel). In addition, the coefficients a_z of these metals are less sensitive to the type of stress.

By contrast, materials with lower strain limits before necking (copper, light alloys) are characterized by high average coefficients $\bar{a}_z$ and are very sensitive to the type of stress.

NOTE.– By placing itself on strain paths such as $\varepsilon_2/\varepsilon_1 = $ const., according to the types of stress:

1) $-0.5 < \varepsilon_2/\varepsilon_1 < 0$ tension;

2) $\varepsilon_2/\varepsilon_1 = 0$ planar strain;

3) $0 < \varepsilon_2/\varepsilon_1 < 1$ expansion;

4) $\varepsilon_2/\varepsilon_1 = 1$ symmetrical biaxial expansion.

For necking Z, we obtain the curves in Figure 4.29 for:

$$a'_z = \left[\frac{\varepsilon_{1z} \text{ (high speed)}}{\varepsilon_{1z} \text{ (low speed)}}\right]_{\varepsilon_2/\varepsilon_1 = C^{te}}$$

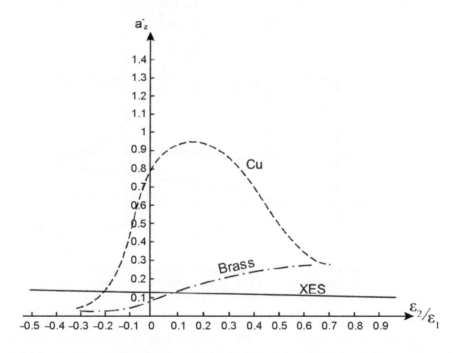

Figure 4.29. *Copper and XES steel (Leroy)*

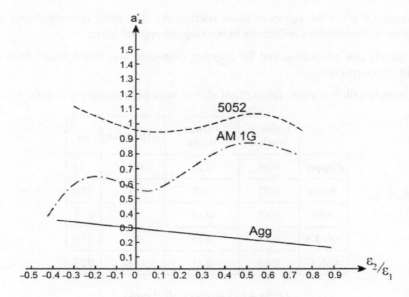

Figure 4.30. *Light alloys (Leroy)*

4.4.8. *Change in fracture*

Figure 4.31 shows the change in R (ε_2) based on the type of strain of the increase in ductility at fracture from previous materials:

$$R\ (\varepsilon_2) = 100 \times \left[\frac{\varepsilon_{1_R}\ (\text{high speed})}{\varepsilon_{1_R}\ (\text{low speed})}\right]_{(\text{at the same } \varepsilon_2)}$$

In the same way as for necking, we define a coefficient a_R characterizing the sensitivity of the ductility in fracture at the strain rate:

$$a_R = \left[\frac{\varepsilon_{1_R}\ (\text{high speed})}{\varepsilon_{1_R}\ (\text{low speed})} - 1\right]_{\varepsilon_2 = \text{const.}}$$

And an average coefficient is given as:

$$\bar{a}_R = \frac{1}{n} \sum_{(\varepsilon_2)}^{n} a_R$$

Table 4.6 gives the values of these coefficients. The same comments that were made for the sensitivity coefficients in necking are applied here:

– metals that are well-suited for pressing (annealed aluminum, brass) have low sensitivity coefficients;

– metals with low strain limits (light alloys) have high sensitivity coefficients.

	Uniaxial tension	Planar strain	Expansion	$\bar{a}_R$
Copper	0.04	0.28	0.28	0.21
Brass	0.02	0.09	0.17	0.09
A99	0.13	0.12	0.10	0.12
AG 2.5	0.88	0.55	0.50	0.60
AM1G	0.60	0.41	0.42	0.54

Table 4.6. *Values of aR (Leroy)*

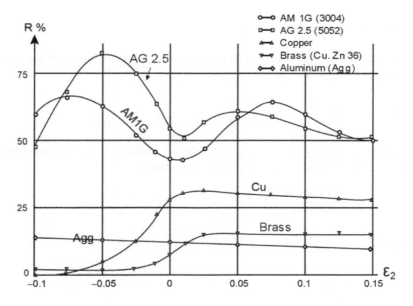

Figure 4.31. *Variation of ductility at fracture, speed sensitivity of strain and according to the type of stress (− 0.1 ⟨ ε_2 ⟨ 0.2)) (Leroy)*

On the other hand, the influence of the type of stress is less clear than during necking. In particular, the significant increase in ductility, which had appeared in the planar strain ($\varepsilon_2 = 0$), no longer appears.

This led us to conduct a more precise study of the field extending from the necking to the fracture.

4.4.9. *Change between necking and fracture*

Figure 4.32 shows the evolution according to the type of stress of the behavior of the materials in the domain extending from the necking to the fracture. The deviation e given as:

$$e = 100 \times \left[\frac{\varepsilon_{1R}^{HV} - \varepsilon_{1Z}^{HV}}{\varepsilon_{1Z}^{HV}} - \frac{\varepsilon_{1R}^{BV} - \varepsilon_{1Z}^{BV}}{\varepsilon_{1Z}^{BV}} \right]_{\varepsilon_2}$$

represents the difference between the percentage of elongation at a high strain rate when the necking begins to the fracture, and the one obtained at low speed.

These curves highlight the differences between the behaviors of materials with high strain limits and materials with lower limits. In fact, for the first ones (A99, brass), no influence of the type of strain can be noticed, and the deviation e remains essentially constant:

e $\simeq$ 0% for brass

e $\simeq$ − 20% for A99

It is interesting to note that this difference is negative in the case of aluminum, which means that at a high strain speed, the domain between necking and fracture is relatively low compared to that at low speed.

For metals with low strain limits (Ag 2.5, AM 1 G), the deviation e is very sensitive to the type of strain and can reach values which are quite high.

On the other hand, it can be noted that this difference decreases rapidly when we approach the planar strain and becomes clearly negative for copper and the AM1G alloy.

$e_{\varepsilon_2} = 0$ (Cu) = − 53%

$e_{\varepsilon_2} = 0$ (AM1G) = − 48%

Therefore, the ductility in necking of these metals increases sharply when they are strained at high speed. On the other hand, after the necking begins, the elongation rate that is still admissible before fracture is clearly influenced by the type of stress.

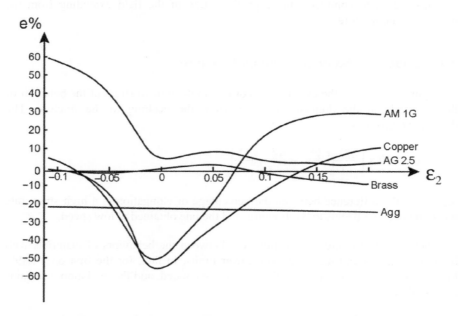

Figure 4.32. *The change of the rates of strain in necking-fracture on the basis of the speed and type of strain (variation of ε_2) (Leroy)*

The electromagnetic or electrohydraulic actions make it possible to obtain high forming speeds $10^2 \ s^{-1} \leq 10^3 \ s^{-1}$, and therefore allow increases in ductility and good metrology due to the speeds of the impacts on the matrix (see examples of the magnetoforming of light alloys, and of the electrohydraulic effect for less conductive metals).

4.4.10. *Examples in magnetoforming and electro-hydroforming*

Taking into account the gains in ductility obtained by the high strain speeds, certain shaping operations can be carried out. The photo shows the case of a tubular expansion operation of hexagonal shape with cutting a piece that cannot be achieved by conventional methods.

In a classic press, the difficulties of carrying out forming without fractures may exist in certain areas. In this case, the introduction of the coils into the areas of the press tools makes it possible to obtain the desired shapes.

Figure 4.33. *Example of an increase in ductility dynamically (Leroy)*

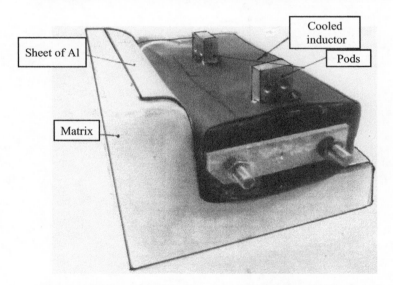

Figure 4.34. *3D inductor for press tools, sheet metal for the body of the automobile (63% increase in zone) (Leroy/Forcem)*

4.4.10.1. *Electrohydraulic*

EXAMPLE 4.8.–

Figure 4.35. *Forming of a stainless steel membrane (Leroy)*

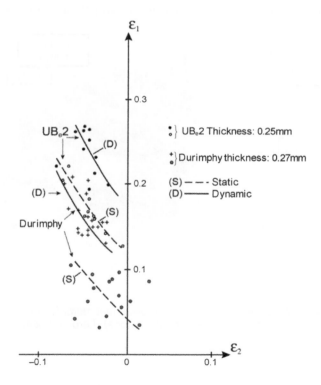

Figure 4.36. *Static and dynamic formability limit curves of two alloys used for membranes in sensors (Leroy/Forcem)*

5

Dynamic Resistance
to Mechanical Shocks

5.1. Shock stresses

5.1.1. *Energy aspects: momentum, kinetic energy, impulse*

A solid S_1 moving at a speed $\vec{v}_1$ (also known as particle speed $\vec{v}_p$) impacts another solid S_2 during the shock, the momentum before and after the shock is conserved.

The momentum of a solid S is:

$$q = m\,v \left| \begin{array}{l} \text{q in kg.m. s}^{-1} \text{ or N.s} \\[4pt] \text{v in m.s}^{-1} \\[4pt] \text{m in kg} \end{array} \right. \qquad [5.1]$$

At the moment of impact, the contact area between the two solids S_1 and S_2 deforms and stores energy, which is then partially or completely restored; a restitution coefficient e is defined as:

$$e = \frac{\text{relative speed of separation}}{\text{relative speed of approach}}$$

That is:

$$e = \frac{v_2' - v_1'}{v_1 - v_2} \left| \begin{array}{l} v_1 > v_2, \text{ speeds of } S_1 \text{ and } S_2 \text{ before shock} \\[4pt] v_1' < v_2' \text{ after shock} \end{array} \right. \qquad [5.2]$$

If:

– e = 1: the shock is elastic (losses = 0);

– 0 < e < 1: applies to most of the shocks;

– e = 0: perfectly plastic shock (maximum losses).

It is necessary to know e before and after the shock $q = C^{te}$, and it is equal to:

$$m_1 v_1 + m_2 v_2 = m_1 v_1' + m_2 v_2' \qquad [5.3]$$

Taking into account the energy losses due to heat exchanges, the plasticity in the strained parts, sonic energy, etc.

In some cases, these losses can be calculated from the variation ΔE_c in kinetic energy before and after the shock.

NOTE.– The kinetic energy of a moving solid is:

$$E_c = \frac{1}{2} m \, v^2 \text{ (in joules)} \qquad [5.4]$$

– For an isolated solid, the work of external forces during a time interval is equal to the change in the kinetic energy of the solid:

$$E_{c_2} - E_{c_1} = [\Delta E_c]_1^2 = [W \, (\Sigma F_{ext})]_{t_1}^{t_2} \qquad [5.5]$$

– For an isolated solid, the derivative of the kinetic energy is equal to the power developed by the external forces:

$$\frac{dE_c}{dt} = P \, (\Sigma F_{ext}) \qquad [5.6]$$

– If the solid or a system isolated from energy (no loss by friction, etc.) and whose forces depend on a potential energy E_p (such as the forces of gravity), if the total energy input E remains constant between two successive instants t_1 and t_2, we obtain:

$$E = [E_c]_1^2 + [E_p]_1^2 = \text{const.}$$

or:

$$E_{c_2} + E_{p2} = E_{c_1} + E_{p_1} = \text{const.} \qquad [5.7]$$

– The result of the external forces ($\Sigma \vec{F}_{ext}$) acting on a solid is equal to the derivative with respect to time of the amount of movement:

$$\Sigma \vec{F}_{ext} = \frac{d\vec{q}}{dt} = \frac{d}{dt}(m\,\vec{v}_G) \text{ (G: center of gravity)} \qquad [5.8]$$

– If we apply a pulse $[I]_1^2$ during an interval of time ($\Delta t = t_2 - t_1$), this pulse is equal to the variation in the momentum during Δt.

$$[I]_1^2 = \vec{q}_2 - \vec{q}_1 = m\vec{v}_2 - m\vec{v}_1 = \int_{t_1}^{t_2} \Sigma \vec{F}_{ext}\, dt \qquad [5.9]$$

The impulse I is expressed in N.s or kg.m. s^{-1}.

NOTE.– When the resultant $\Sigma \vec{F}_{ext}$ is a constant over time, the expression can be simplified to:

$$\int_{t_1}^{t_2} \Sigma \vec{F}_{ext}\, dt = \Sigma \vec{F}_{ext} \int_{t_1}^{t_2} dt = \Sigma \vec{F}_{ext}\,(t_2 - t_1)$$

$$= \Sigma \vec{F}_{ext}\, \Delta t = [I]_1^2 \qquad [5.10]$$

5.1.2. *Comparison between stress levels in static and dynamic loads*

5.1.2.1. *Movements and stresses*

We will use σ_s and δ_s to refer to the static stresses and displacements due to external forces F_s applied slowly, and similarly, σ_d and δ_d dynamically for external forces F_d applied quickly.

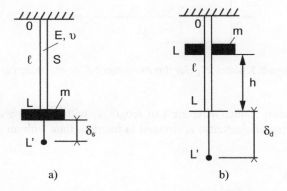

a)   b)

Figure 5.1. *Elongation δ_s and δ_d statically and dynamically from L to L'.*
(a) Static force. (b) Dynamic force from the fall of the mass m of h

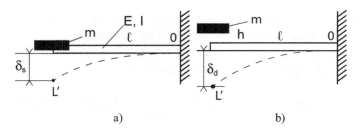

Figure 5.2. *Arrows δ_s and δ_d flexed statically and dynamically. (a) Static load. (b) Dynamic load from the fall of the mass m from h*

5.1.2.2. *Static traction (case a)*

The movement δ is proportional to the force F. For an elementary displacement $\Delta\delta$, the work of the force $F(\delta)$ is $F(\delta) \times \Delta\delta$ and, for a finite displacement δ, the work W of the force is equal to the sum of the elementary areas.

$$W_{L'} = \frac{(\delta \times F) L'}{2} \text{ (area/work)}$$ [5.11]

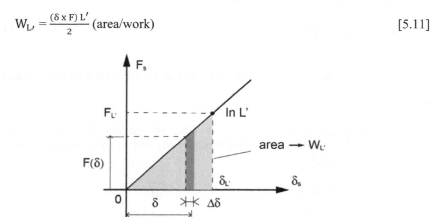

Figure 5.3. *Work $W_{L'}$ for the movement δ_s in elasticity (a)*

The object (beam, cable, wire, etc.) of length ℓ, with section S and of elastic stiffness E (modulus of elasticity) is stressed in traction, thus with an elongation $\Delta\ell$ with:

$$\sigma = \frac{F}{S} = \frac{\Delta\ell}{\ell} E$$

We obtain:

$$\Delta\ell = \frac{F}{S}\frac{\ell}{E} = \delta_{L'} \qquad [5.12]$$

– [5.11]: W of L to L' = $\delta \times$ F/2   with [5.12]:

$$W_s = \frac{F^2\ell}{2\,S\,E} \qquad [5.13]$$

or as a function of δ:

$$W_s = \frac{S\,E\,\delta^2}{2\,\ell} \qquad [5.14]$$

– [5.13] and [5.14]: work for static strain;

– [5.12]: movement and static stress.

$$\delta_s = \frac{F\ell}{S\,E} \text{ and } \sigma_s = \frac{F}{S}$$

5.1.2.3. Dynamic traction (case b)

The mass m is in motion with a fall of h:

– at t_0, the mass is motionless in L;

– at t_1, the mass is in L', with zero speed at δ_d:

$$W_{kinetic}\left.\right|_{t_0}^{t_1} = 0 \text{ because the speed is zero at } t_0 \text{ and } t_1$$

$$W_{potential}\left.\right|_{t_0}^{t_1} = W_{elastic\,force}\left.\right|_{0}^{t_1}$$

It gives:

$$mg\,(h + \delta_d) = S\,E\,\delta_d^2/2\ell \qquad [5.15]$$

and thus:

$$S\,E\,\delta_d^2 - 2\,mg\,\ell\,\delta_d - 2\,mg\,\ell\,h = 0 \qquad [5.16]$$

Here, the roots are:

$$\delta_d = \frac{A \pm \sqrt{A^2 + 2\,S\,E\,A\,h}}{2\,S\,E} \qquad [5.17]$$

By placing m g ℓ (or m γ l) = A.

Now, as we have seen in [5.12] statically, the movement δ_s = mg ℓ/S E = A/SE and then when taken together with [5.17]:

$$\delta_d = \delta_s (1 + \sqrt{1 + 2 h/\delta_s})$$ [5.18]

In the case where h = 0, we have $\delta_d = 2 \delta_s$ for the application of a force instantly (or almost instantly) on the part and without any shock; for example, this is the case for lifting a load.

5.1.2.4. Maximum dynamic stress

Dynamic stress [5.12]:

$$\sigma_d = \frac{\delta_d}{\ell} E$$

with [5.18], we obtain:

$$\sigma_d = \sigma_s (1 + \sqrt{1 + 2 hE/\sigma_s\ell})$$ [5.19]

In conclusion, for h = 0 with a finite force applied abruptly:

$$\delta_d \simeq 2 \delta_s \text{ and } \sigma_d \simeq 2 \sigma_s$$ [5.20]

EXAMPLE. 5.1.–

A mass of 50 kg falls by h = 10 cm and tenses a rod made of Al 2014 T4 with an elastic limit σ_0 = 280 MPa, E = 70 GPa, the rod has a length ℓ = 1 m and a diameter $\emptyset$ = 4 mm. We determine its dynamic elongation and under the same force in static.

SOLUTION 5.1.–

In static:

[5.12] $\delta_s = F\ell / S E$

$\delta_s = (50 \times 9.81 \times 1)/\pi \times 2^2 \times 10^{-6} \times 7 \times 10^{10}$

$\simeq 0.56$ mm

$\sigma_s = 50 \times 9.81/\pi \times 2^2 \times 10^{-6} = 39$ MPa

In dynamic:

(see equation [5.18]) $\delta_d = \delta_s \left(1 + \sqrt{1 + 2\,h/\delta_s}\right)$

$\delta_d = 0.56 \left(1 + \sqrt{1 + 2 \times 100 \times 10^{-3}/0.56 \times 10^{-3}}\right)$

$\quad = 11.15$ mm

(see equation [5.19]) $\sigma_d = \sigma_s \left(1 + \sqrt{1 + 2\,h\,E/\sigma_s\,\ell}\right)$

$\quad = 39 \left(1 + \sqrt{1 + 2 \times 100 \times 10^{-3} \times 7 \times 10^{10}/39 \times 10^6 \times 1}\right)$

$\quad = 780$ MPa, that is, $\sigma_d \simeq 20\,\sigma_s$

– Value $\gg \sigma_0$ and R_m: fracture. If $h = 0$: $\delta_d = 2 \times 0.56 = 1.12$ mm:

[5.10] $\sigma_d = 2 \times 39 = 78$ MPa $< \sigma_0$ (elasticity)

5.1.2.5. Static flexing (case a)

The work of elastic deformation in flexing W_f is proportional to Mf^2 (the square of the flexing moment) and inversely proportional to the elastic stiffness E of the material, as well as to the values of I of the sections.

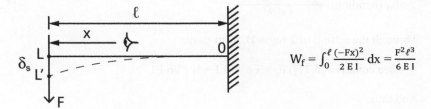

$$W_f = \int_0^\ell \frac{(-Fx)^2}{2\,E\,I}\,dx = \frac{F^2\ell^3}{6\,E\,I}$$

Figure 5.4. *Static deflection δs*

Deflection by application of Castigliano's theorem:

$$\delta_s = \frac{\partial W_f}{\partial F} = \frac{F\ell^3}{3\,E\,I} \tag{5.21}$$

5.1.2.6. Dynamic flexing (case b)[1]

$W_{Fext} = W_{Fint}$ with $W_{Fint} = (F_d \times \delta_d)/2$

---

1 Figure 5.5.

We obtain $m_g (h + \delta_d) = (F_d \times \delta_d)/2$ giving:

$$F_d = 2 \, mg \, (h + \delta_d)/\delta_d \qquad\qquad [5.22]$$

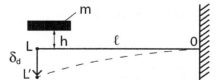

The external work is linked to the displacement of the force F

$$W_{Fext} = mg \, (h + \delta_d) \qquad [5.23]$$

Figure 5.5. *Dynamic deflection δd*

According to [5.21] and [5.22]:

$$2mg \, (h + \delta_d)/ \, \delta_d = F_d = 3 \, E \, I \, \delta_d/\ell^3$$

That is:

$$\frac{3\,E\,I}{\ell^3} \, \delta_d^2 - 2mg \, \delta_d - 2mg \, h = 0 \qquad\qquad [5.24]$$

Here the roots are given as:

$$\delta_d \text{ (bending) } \frac{B \pm \sqrt{B^2 + 12 \, EI \, B \, h/\ell^3}}{6 \, EI/\ell^3} \qquad\qquad [5.25]$$

Through the setting of $2 \, mg = B$, or in static:

(see equation [5.21]) $\delta_s = F\ell^3/3 \, EI = B \, \ell^3/6 \, EI$

And thus:

$$\delta_d \text{ (bend)} = \delta_s + \sqrt{\delta_s^2 + 2 \, h \, \delta_s} \qquad\qquad [5.26]$$

EXAMPLE 5.2.–

A force $F = 10$ N is applied to the end of a beam of length $\ell = 1$ m and of a rectangular section (in mm) 10×60. The beam is recessed at 0 (see Figures 5.4 and 5.5) and undergoes static or dynamic bending (height drop h of the mass), beam material: Al 2014 T4 ($\sigma_0 = 280$ MPa, $E = 70$ GPa).

Determine:

– in static conditions, the values of the two arrows δ_s according to section-force positions;

– in dynamic conditions, the two values of drop height h in elastic limit of the beam according to the positions of the section.

SOLUTION 5.2.–

By calling for the section a × b (b = 6a).

– Two cases for I:

$$I_{(1)} = \frac{ba^3}{12} = \frac{a^4}{2} = \frac{10^{-8}}{2} \ m^4$$

$$I_{(2)} = \frac{ab^3}{12} = 18 \ a^4 = 18 \times 10^{-8} \ m^4$$

$$I_{(2)} = 36 \ I_{(1)}$$

$$[5.21] \ \delta_{s(1)} = \frac{F\ell^3}{3 \ E \ I} = \frac{10 \times 1^3 \times 2}{3 \times 70 \times 10^9 \times 10^{-8}} \Rightarrow 9.5 \ mm$$

$$\delta_{s(2)} = \delta_{s(1)}/36 = 0.26 \ mm$$

– In bending, the stresses are the strongest at 0 (embedding) or $M_{bending} = F_d \times 1$ with F_d in L', giving a bending arrow δ_d equal to $\delta_d = F_d \ \ell^3/3rd \ I$, from which $M_{bending} = 3 \ E \ I \ \delta_d/\ell^2$ (Nm) and $\sigma_{max} = \frac{M_f}{I} \times v$ with $v = a/2$ or $3a$ in elastic limit $\sigma_{max} = \sigma_0$ (Re), we have $\sigma_0 = (3 \ E \ I \ \delta_d/\ell^2) \times (v \ / \ I)$, give δ_d equal to, respectively:

$$\delta_{d(1)} = \frac{\sigma_0 \ell^2}{3 \ E \ v_{(1)}} \ and \ \delta_{d(2)} = \frac{\sigma_0 \ell^2}{E \ v_{(2)}}$$

$$\delta_{d(1)} = \frac{280 \times 10^6 \times 1^2}{3.70 \times 10^9 \times 5 \times 10^{-3}} \rightarrow 266 \ mm$$

$$\delta_{d(2)} = \frac{280 \times 10^6 \times 1^2}{3.70 \times 10^9 \times 30 \times 10^{-3}} \rightarrow 44 \ mm$$

Equation [5.26] gives $h = \delta_d \ (\delta_d - 2\delta_s)/2\delta_s$.

5.2. Resilience test

5.2.1. *Impact by a simple pendulum*

A pendulum of weight $\overrightarrow{mg}$ is released without any initial speed from a point B of altitude z_1 to produce an impact on O. Knowing that the only force producing work is the weight (there is no air resistance nor friction in A) and that the tension $\overrightarrow{T}$ from the cable of length ℓ does not produce any, determine the speed $\vec{V}$ at the point M of altitude z_2 (Figure 5.6).

The pendulum does not exchange energy (being an energy isolated system), and we have:

(see equation [5.7]) $E_{C_M} + E_{P_M} = E_{C_B} + E_{P_B}$

from which we have $\frac{m\,V_M^2}{2} + mg\,z_2 = \frac{mV_B^2}{2} + mg\,z_1$ with $V_B = 0$.

And $z_1 - z_2 = h$, we obtain:

$$\frac{m\,V_M^2}{2} = mg\,(z_1 - z_2)$$

and $V_M = \sqrt{2\,g\,h}$.   [5.27]

At point 0 of impact if:

$$h = \ell,\ V_{impact} = \sqrt{2\,g\,\ell}$$ [5.28]

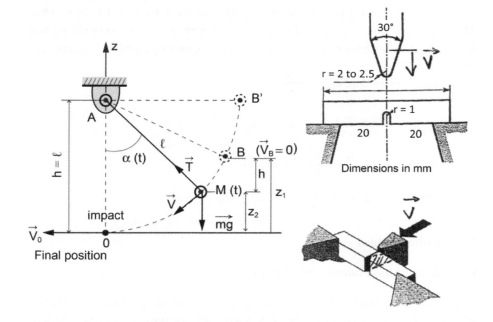

Figure 5.6. *Impact of pendulum on test piece*

By placing a test sample at 0, the energy absorbed in the impact is obtained relative to a section of unit area of the material tested. This fracture energy is called *resilience*. It is obtained by the Charpy pendulum model (Figure 5.7) and corresponds to h, and it is of great interest for the study of the transitions between brittle and ductile ruptures of materials.

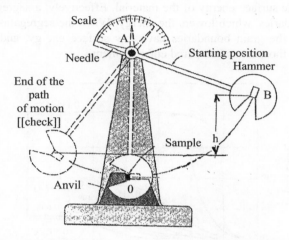

Figure 5.7. *Charpy's sheep pendulum for resilience test of materials*

5.2.1.1. *Brittle-to-ductile transition*

The energy of the fracture of certain materials (particularly CC metals) varies considerably with temperature. These materials have a brittle-to-ductile transition.

At low temperatures, the propagation of the crack remains brittle in nature. At a sufficiently high temperature, it becomes ductile. Then a transition occurs from one propagation mode to another, and the brittle or ductile behavior of this type of material is determined by the temperature at which the test is carried out. The transition temperature is characteristic of the material under the conditions of the test.

The impact behavior of the materials also depends on the crystal structure. In this way, CFC materials never seem to exhibit brittle behavior, unlike CC and HCP materials.

– V, Nb and Ta (CC) experience a brittle-to-ductile transition at roughly 150K or less.

– W, Mo and Cr are more likely to exhibit brittle behavior than the materials listed above, and as such, they have higher transition temperatures.

The brittle fracture tends to be intergranular when the degree of purity of the material decreases. The important parameter in determining the form of the brittle fracture is thus the surface energy of the material. Effectively, a segregation occurs at the grain boundaries, which lowers the energy. When the segregation occurs at a significant level, the grain boundaries have low surface energy, and the fracture tends to spread in those areas (tempered embrittlement).

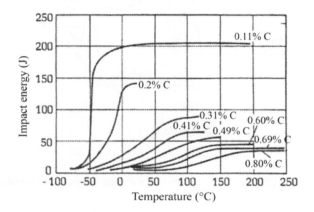

Figure 5.8. *Temperature resilience of carbon steels*

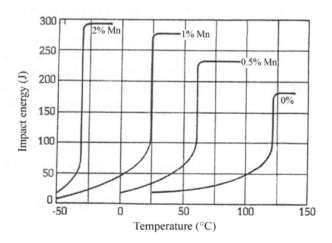

Figure 5.9. *Temperature resilience of Fe-Mn alloys-0.05% C*

The brittle behavior of the CCS is facilitated by the presence of impurities capable of forming interstitial solid solutions. A classic example is carbon in iron.

EXAMPLE 5.3.–

Use of a pendulum: ballistic test cases.

The objective is to determine the velocity v of a projectile using a pendulum made up of a sandbag with a mass M = 30 kg, suspended from a thin rope of length ℓ = 2 m. A projectile of a mass m = 60 g swings the pendulum to an incline of an angle θ of 20°. We determine the velocity v of the projectile.

SOLUTION 5.3.–

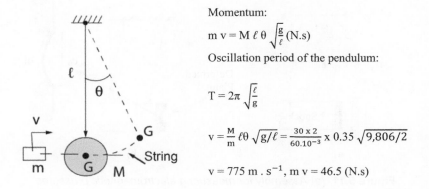

Momentum:

$$m \, v = M \, \ell \, \theta \, \sqrt{\frac{g}{\ell}} \, (N.s)$$

Oscillation period of the pendulum:

$$T = 2\pi \sqrt{\frac{\ell}{g}}$$

$$v = \frac{M}{m} \, \ell\theta \, \sqrt{g/\ell} = \frac{30 \times 2}{60.10^{-3}} \times 0.35 \, \sqrt{9,806/2}$$

$$v = 775 \, m \cdot s^{-1}, \, m \, v = 46.5 \, (N.s)$$

Figure 5.10. *Impact on the pendulum*

5.2.1.2. *Internal "electronic" shock by electromagnetic pulse*

Shocks can be obtained in electrically conductive materials by applying an intense pulsed magnetic field.

The electronic action (the Laplace force) causes a stress wave within the thickness of the skin of the material, and it acquires kinetic energies that are favorable to industrial applications (forming, welding, assembly, etc.). The stress wave is proportional to the square of the magnetic field applied to the conductive material. In order to determine the pressures that are exerted on the material, we used a pendulum measuring the shock pulse on various materials (Al, Cu, etc.) stressed by the action of plane inductors (spiral plating coil in volume II; Revue *Phys. Appl* 36 and *French Journal of Mechanics*, no. 77).

Under the effects of the electromagnetic pressure P(t) applied over the span of a few microseconds, the mass M of the pendulum reaches a speed v_0 and, with an active surface S of the inductor, we obtain:

(see equation [5.19]) $[I]_0^\infty = S \int_0^\infty P(t) \, dt = M \, v_0$

and thus a (measurable) movement of the pendulum and the pressure, which are produced by mounting the test sample (metal plate) on a solid and rigid insulating matrix.

The stamping deformation tests of the test pieces are carried out with the same type of device using a hollow matrix (see Figure 5.11 for both cases).

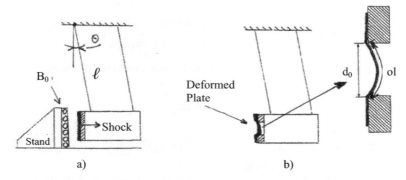

Figure 5.11. *(a) Assembly for measuring electromagnetic pressures (electronic repulsion), internal shock to the material (no tools used). (b) High-speed stamping of hollow-matrix deformation*

5.2.2. Stress from polar shock impacts

5.2.2.1. Impact stress

The stress wave for a material with a volumetric mass ϱ acoustic velocity c and particle velocity V_p is written as:

$$\sigma = \varrho \, c \, V_p = Z \, V_p$$

Z is the acoustic impedance of the medium. We assume a part impacting a matrix (Figure 5.12).

Before the impact:

– medium $1 = V_p = \dot{d}$;

– medium $0 = V_p = 0$.

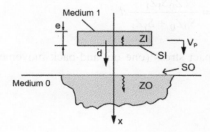

Figure 5.12. *Impact*

At the time of the impact ($t = 0$), the forces generated will be of the same value on both mediums.

$$F_0 = S_0\, Z_0\, V_p + A$$

However, just before the impact:

$$S_0\, Z_0\, V_p = 0 \Rightarrow A = 0 \text{ and: } F_1 = -\, S_1 Z_1 v_p + B$$

Similarly, just before the impact, there is no stress:

$$F_1 = 0 \Rightarrow B = S_1 Z_1 \dot{d}$$

$$A\, t = 0: F_0 = F_1 = F \Rightarrow S_0 Z_0 v_p = -\, S_1 Z_1 v_p + S_1 Z_1 \dot{d}$$

And thus the particle velocity at impact is:

$$v_\rho = \frac{S_1 Z_1}{S_0 Z_0 + S_1 Z_1}\, \dot{d}$$

The impact force can be written as:

$$F = \frac{S_0 Z_0\, S_1 Z_1}{S_0 Z_0 + S_1 Z_1}\, \dot{d}$$

The shock stresses from the sample to the matrix are:

$$\sigma\ sample = \frac{F}{S_1} = \frac{S_0 Z_0 Z_1}{S_0 Z_0 + S_1 Z_1} \dot{d}$$

$$\sigma\ matrix = \frac{F}{S_0} = \frac{Z_0 S_1 Z_1}{S_0 Z_0 + S_1 Z_1} \dot{d}$$

Duration of the impact stress (one out-and-back movement of the wave in the sample) is given as:

$$t = \frac{2e}{C_1}$$

In the case that $S_1 = S_0 = S$, the impact force becomes:

$$F_{impact} = \frac{S Z_0 Z_1}{Z_0 + Z_1} \dot{d}$$

and the impact stress becomes:

$$\sigma = \frac{Z_0 Z_1}{Z_0 + Z_1} \dot{d}$$

5.2.2.2. Polar shock

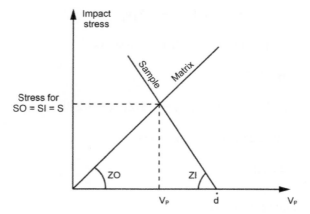

Figure 5.13. *Polar shock*

EXAMPLE 5.4.–

– For a 1050A aluminum:

- longitudinal impedance $Z = 1.36 \times 10^7$ kg/m^2.s;

- velocity of the c waves = 5,055 m/s.

– For steel:

- longitudinal impedance $Z = 4 \times 10^7$ kg/m^2.s;

- velocity of the c waves = 5,064 m/s.

5.2.2.3. *Impact values*

The impact of *aluminum on aluminum* at $\dot{d} = 1$ m/s:

$$\sigma_{impact} = \frac{1.36 \; 10^7 \; 1.36 \; 10^7}{(1.36 + 1.36)10^7} \; 1 = 6.8 \text{ MPa}$$

The impact of *aluminum on steel* at $\dot{d} = 1$ m/s:

$$\sigma_{impact} = \frac{4 \; 10^7 \; 1.36 \; 10^7}{(4 + 1.36)10^7} \; 1 = 10.14 \text{ MPa}$$

The duration of the impact stress (out and back) for an aluminum sheet with a thickness of 1 mm is given as:

$$t = \frac{2 \times 1.10^3}{5,055} = 0.39 \; \mu s$$

5.2.2.4. *Flaking*

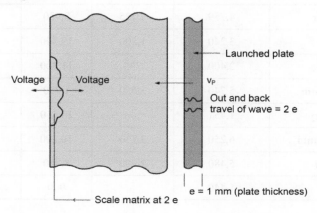

Figure 5.14. *Example of successive flaking of a steel matrix by internal waves (the case of impacts of Al on steel)*

Acoustic velocities, acoustic impedances, and densities (for materials in common use)			Z Kg m^{-2} s^{-1}
Material	**Compression speed v_e (m/s) C_0**	**Shear speed v_s (m/s)**	**Density ρ (kg/m^3)**	**Specific acoustic impedance $\rho v_e \times 10^6$**
Air	332	–	1.205	0.0004
Aluminum	6,400	3,130	2,700	17.3
Beryllium	12,890	8,880	1,800	23.2
Brass 70/30	4,372	2,100	8,450	37.0
Cast iron	3,500	2,200	7,200	25
	5,600	3,200	–	40
Copper	4,759	2,325	8,930	42.5
Gold	3,240	1,200	19,300	63
Lead	2,400	790	11,300	27.2
Magnesium	5,740	3,080	1,720	9.9
Mercury	1,451	–	13,550	19.6
Molybdenum	6,250	3,350	10,200	63.7
Nickel	5,480	2,990	8,850	48.5
Oil	1,440	–	900	1.3

Perspex	2,680	1,320	1,200	3.2
Platinum	3,960	1,670	21,400	85.0
Polystyrene	2,350	1,120	1,060	2.5
Mild steel	5,960	3,240	7,850	46.7
Stainless steel	5,740	3,130	7,800	44.8
Silver	3,704	1,698	10,500	36.9
Tin	3,380	1,609	7,300	24.7
Titanium	5,990	3,120	4,500	27,0
Tungsten	5,174	2,880	19,300	100.0
Tungsten carbide	6,655	3,984	10,000	66.5
Uranium	3,370	2,020	15,000 / 18,700	98.5 / 63.0
Water	1,480	−	1,000	1.48
Zinc	4,170	2,480	7,100	29.6
Zirconium	4,650	2,300	6,400	29.8

Table 5.1. *Impedance of materials (data are indicative)*

NOTE.– These values were taken from the "Tables of physical and chemical constants". Using these values and comparing them with those obtained from other sources, it must be recalled that they are dependent on the state of the material (melted, annealed, quenched, etc.). For all common purposes, the above figures are largely sufficient in terms of accuracy.

5.2.3. *Shock with rebound, coefficient of restitution, energy losses*

A block A (hammer) of a mass M is used to impact the piece B (causing depression, crushing, etc.).

From a resting position, it falls from a height h onto B and bounces to a height r (the rebound):

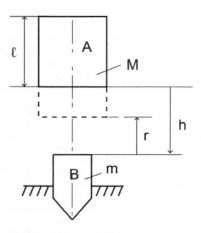

Figure 5.15. *Impact with rebound r*

– v_1 is the speed of A just before impact, where we have:

$$1/2 \, Mv_1^2 = M \, g \, h \text{ from where } v_1 = \sqrt{2 \, g \, h} \tag{5.29}$$

– The energy is conserved on the descent and ascent and, just after impact, the speed v_1' of A is equal to:

$$1/2 \, Mv_1'^2 = M \, g \, r \text{ and thus } v_1' = \sqrt{2 \, g \, r} \tag{5.30}$$

– For the grouping A + B, the momentum is conserved:

$$mv_2 + Mv_1 = mv_2' + Mv_1' \tag{5.31}$$

with $v_2 = 0$ just before impact, thus:

$$v_2' = \frac{M}{m} (v_1 - v_1'), \, v_1' \text{ negative} \tag{5.32}$$

EXAMPLE 5.5.–

Given that h = 2 m, M = 100 kg, m = 200 kg, r = 10 cm, v_1 = 6.26 ms^{-1}, v_1' = 1.4 ms^{-1}, v_2' = 3.83 ms^{-1}.

– Coefficient of restitution obtained:

$$e = \frac{v_2' - v_1'}{v_1 - v_2} = \frac{3.83 - (-1.4)}{6.26 - 0} = 0.83 \tag{5.33}$$

– Energy losses: kinetic energies before Ec_1 and after impact Ec_2:

$$Ec_1 = \frac{1}{2} Mv_1^2 = \frac{1}{2} Mgh = 1,959 \text{ joules}$$

$$Ec_2 = \frac{1}{2} Mv_1'^2 + \frac{1}{2} mv_2'^2 = 98 + 1,467 = 1,565 \text{ D}$$

– Percentage of energy loss:

$$\frac{Ec_1 - Ec_2}{Ec_1} \times 100 = 20\%$$

– Duration of the impact: $\Delta t = 2 \ \ell/c$; C: acoustic velocity of A ($\simeq 5,000 \text{ ms}^{-1}$/steel).

5.2.4. *Effect of resistance to movement, speed and stress upon impact*

The case of a fall without resistance and no initial speed at t = 0 is expressed as:

$$mg = m \frac{dv}{dt} \text{ and thus } gdt = dv$$

It gives:

$$v = \int_0^t gdt = gt \rightarrow \text{ in 1 sec: } v \simeq 9.8 \text{ ms}^{-1} \qquad [5.34]$$

Movement z:

$$\frac{dz}{dt} = v = gt \rightarrow dz = gt \cdot dt$$

$$z = \int_0^t gt.dt = \frac{1}{2} gt^2 \rightarrow \text{ in 1 s: } z \simeq 5 \text{ m} \qquad [5.35]$$

5.2.4.1. *Resistance force proportional to the instantaneous speed v*

– Resistance force $f = - b \, m \, v$:

$$(M L T^{-2}/M L T^{-1} \rightarrow b = C^{te} \text{ in } s^{-1}) \qquad [5.36]$$

– Speed at the time t:

$$mg - b \, m \, v = m \frac{dv}{dt} \rightarrow g - bv = \frac{dv}{dt} \qquad [5.37]$$

and thus $\frac{dv}{g - bv} = dt$ giving $-\frac{1}{b} \ln (g - bv) = t + C^{te}$ where for t = 0, $v = v_0$, we have:

$$\ln \frac{g - bv}{g - bv_0} = - bt$$

or:

$$g - bv = (g - bv_0)\, e^{-bt} \tag{5.38}$$

– Instantaneous speed of impact at time t:

$$v = \frac{g}{b} - \left(\frac{g}{b} - v_0\right) e^{-bt} \tag{5.39}$$

– The case where the initial speed $v_0 = 0$:

$$v = \frac{g}{b}\left(1 - e^{-bt}\right) \tag{5.40}$$

5.2.4.2. Limiting speed

When the acceleration $\frac{dv}{dt}$ approaches zero: [5.37] $g - bv = 0$, the limiting speed v_ℓ is therefore:

$$v_\ell = \frac{g}{b} \tag{5.41}$$

or again, with [5.38] for $t \to \infty$, we obtain:

$$(g - bv_0)\, e^{-\infty} = 0 \to g - bv = 0 \text{ value in [5.41]} \tag{5.42}$$

The distance z traveled is obtained with [5.42]:

$$\frac{dz}{dt} = v = v_\ell - (v_\ell - v_0)\, e^{-bt}$$

by integration:

$$z = v_\ell t + \frac{v_\ell - v_0}{b}\, e^{-bt} + z_0$$

with $t = 0$, $z = 0$, and thus:

$$z_0 = \frac{-(v_\ell - v_0)}{b}\, e^{-bx0},\ e^{\circ} = 1$$

giving the distance z traveled:

$$z = v_\ell t - \frac{v_\ell - v_0}{b}\left(1 - e^{-bt}\right) \tag{5.43}$$

and $v_\ell = \frac{g}{b}$ (limiting speed) $\hfill$ [5.44]

$v = v_\ell - (v_\ell - v_0) \, e^{-bt}$ (v: instantaneous speed) $\hfill$ [5.45]

During the impact, at the time t_{choc} and with an initial velocity of zero ($v_0 = 0$), the momentum of a mass m is equal to:

$$q = m \, v_\ell \, (1 - e^{-bt}) \hfill [5.46]$$

5.2.4.3. *Acceleration*

According to equation [5.39]: $v = \frac{g}{b} - \left(\frac{g}{b} - v_0 \right) e^{-bt}$.

Acceleration $\gamma = \frac{dv}{dt} = - b \left(\frac{g}{b} - v_0 \right) e^{-bt}$:

$$\gamma = (g - bv_0) \, e^{-bt} \hfill [5.47]$$

If the initial speed $v_0 = 0$:

$$\gamma = g \, e^{-bt} \hfill [5.48]$$

for $t = 0 \rightarrow \gamma = g$.

$$t \rightarrow \infty \rightarrow \gamma = 0$$

NOTE.– Introduction of the limiting speed v_ℓ in the expression of γ.

We obtain:

$$e^{-bt} = \frac{v_\ell - v}{v_\ell - v_0} \hfill [5.49]$$

$$\gamma = b \, (v_\ell - v_0) \, \frac{v_\ell - v}{v_\ell - v_0}$$

That is:

$$\gamma = b \, (v_\ell - v) \hfill [5.50]$$

Duration t of the movement until the impact is given as:

$$t = \frac{1}{b} \ln \frac{v_\ell - v_0}{v_\ell - v} \hfill [5.51]$$

EXAMPLE 5.6.–

Calculation of the time t needed for the speed to reach 0.999 of the limiting speed v_ℓ.

In the case of t = 0, $v_0 = 0$, we have with [5.37] $g - bv = \frac{dv}{dt}$ to be multiplied by b, which gives:

$$\frac{-dv}{g/b - v} = - b\, dt \text{ with } \frac{g}{b} = v_\ell$$

by integration:

$$\ln \frac{v_\ell - v}{v_\ell} = - bt \text{ and for } v = 0.999\, v_\ell$$

$$\ln 10^{-3} = - bt$$

thus: $t = \frac{3 \ln 10}{b}$.

If b = 25 s^{-1}, we obtain: t $\simeq$ 0.3 s.

5.2.5. *Resistant force proportional to the square of the instantaneous speed v²*

The resistance force f depends on v^2, and we obtain:

$$f = - k\, m\, v^2 \tag{5.52}$$

– Dimension of k: L^{-1}:

$$F(N): L\, M\, T^{-2}, m\, v^2: M\, L^2\, T^{-2}$$

$$mg - k\, m\, v^2 = m \frac{dv}{dt} \tag{5.53}$$

$$g - k\, v^2 = \frac{dv}{dt} \tag{5.54}$$

We set:

$$g = k\, \lambda^2 \tag{5.55}$$

– Dimension of λ: LT^{-1} (speed):

(see equation [5.54]) $k (\lambda^2 - v^2) = \frac{dv}{dt}$

$$\frac{dv}{\lambda^2 - v^2} = k \, dt \qquad\qquad [5.56]$$

– Integral of the form $\frac{1}{a^2 - x^2}$: $\rightarrow \frac{1}{a}$ Arg th $\frac{x}{a}$ + const. thus:

$$\frac{1}{\lambda} \text{ Arg th } \frac{v}{\lambda} = kt + \text{const.} \qquad\qquad [5.57]$$

Arg th $\frac{v}{\lambda} = \lambda \, kt + \lambda.\text{const.}$

and $\frac{v}{\lambda} =$ th $(\lambda \, kt + \lambda.\text{const.})$ $\qquad\qquad [5.58]$

– The case where: if at $t = 0$, $v = 0$:

(see equation [5.58]) $0 = \lambda$ th $(0 + \lambda.\text{const.})$ and $\lambda.\text{const.} = 0$

and thus:

$$v = \lambda \text{ th } (\lambda \, kt) \qquad\qquad [5.59]$$

with the speed obtained when $\gamma = 0$ or for $t \rightarrow \infty$:

(see equation [5.54]) $\frac{dv}{dt} = 0$ for $\lambda = v_\ell$ $\qquad\qquad [5.60]$

(see equation [5.59]) $v = v_\ell$ th $(v_\ell \, k \, t)$ $\qquad\qquad [5.61]$

(see equation [5.55]) $v_\ell = \sqrt{\frac{g}{k}}$ $\qquad\qquad [5.62]$

NOTE.– th $x = \frac{1 - e^{-2x}}{1 + e^{-2x}}$

The instantaneous speed v can also be given if it is known that:

$k = g/\lambda^2$ and that $\lambda = v_\ell$

(see equation [5.61]) $v = v_\ell$ th $\left(\frac{g}{v_\ell} t\right)$ $\qquad\qquad [5.63]$

EXAMPLE 5.7.–

If the limiting velocity of a solid is $v_\ell = 40$ m s$^{-1} = \lambda$.

We obtain: $\frac{g}{v_\ell} \simeq \frac{10}{40} = 0.25$ s^{-1} and $v = 40$ th $(0.25.t)$.

$$k = g/\lambda^2 = g/v_\ell^2 = 6.25 \times 10^{-3} \text{ m}^{-1}$$

and thus $F_{resistance} = -6.25 \times 10^{-3}$ m v^2.

Path traveled:

$$z = \int_{t=0}^{t} v.\, dt \text{ with [5.63], } v = v_\ell \text{ th } (\tfrac{g}{v_\ell}.t)$$

– Recall:

$$\int \text{th } x.\, dx = \ln \text{ ch } x + C^{te} \text{ and ch } x = \frac{e^x + e^{-x}}{2}$$

with $x = \frac{g}{v_\ell t}$, we have: $dx = \frac{g}{v_\ell} dt$ and thus:

$$z = \frac{v_\ell}{g/v_\ell} \int_0^t \text{th } (\tfrac{g}{v_\ell}.t).\tfrac{g}{v_\ell} dt$$

$$z = \frac{v_\ell^2}{g} \left[\ln \text{ ch } \tfrac{g}{v_\ell} t \right]_0^t$$

$$z = \frac{v_\ell^2}{g} \ln \text{ ch } \tfrac{g}{v_\ell} t \text{ (knowing that } \tfrac{g}{v_\ell} t \text{ is a number)}$$

or:

$$z = \frac{v_\ell^2}{g} \ln \frac{e^{\frac{g}{v_\ell}t} + e^{\frac{-g}{v_\ell}t}}{2} \tag{5.64}$$

EXAMPLE 5.8.–

Let us return to the previous case of speed limit, $v_\ell = 40$ m s^{-1}, and calculate the path z traveled in 2 s:

(see equation [5.64]) $z = \frac{40}{0.25} \ln \text{ ch } 2 \times 0.25 \text{ car } g/v_\ell \simeq 0.25$

$$z = 160 \ln \text{ch } 0.5 = 160 \ln \frac{e^{0.5} + e^{-0.5}}{2}$$

$$= 160 \ln 1.127 = 19.14 \text{ m}$$

NOTE. If there is no resistance force:

$$z = \frac{1}{2} g t^2 \simeq 20 \text{ m}$$

– Acceleration of the solid and acting force, calculation of γ, knowing that:

(see equation [5.63]) $v = v_\ell \text{ th} \left(\frac{g}{v_\ell}t\right)$ with th $x = \frac{1 - e^{-2x}}{1 + e^{-2x}}$

$$v = v_\ell \times (1 - e^{-2gt/v\ell})/(1 + e^{-2gt/v\ell}) \qquad\qquad [5.65]$$

$$\gamma = \frac{dv}{dt} = 4 \frac{g}{v_\ell} v_\ell \frac{e - \frac{eg}{v_\ell} t}{(1 + e - \frac{2g}{v_\ell} t)^2} \qquad\qquad [5.66]$$

NOTE. At the beginning of the movement:

$$t = 0 \text{ and } e^{-\frac{2g}{v\ell} t} = 1$$

$$\gamma = 4g \frac{1}{(1 + 1)^2} = g$$

– Resultant force on the solid S: $F = m \gamma$ knowing that:

$$e^{-\frac{2gt}{v\ell}} = \frac{v_\ell - v}{v_\ell + v}$$

We obtain:

$$F = 4 \, mg \frac{(V_\ell - v)/(V_\ell + v)}{\left(1 + \frac{V_\ell - V}{V_\ell + V}\right)^2} \qquad\qquad [5.67]$$

giving $F = 4 \, mg \frac{v_\ell^2 - v^2}{4 v_\ell^2}$.

And thus:

$$F = mg \left(1 - \frac{v^2}{v_\ell^2}\right) \qquad\qquad [5.68]$$

$$= mg - mg \left(\frac{v}{v_\ell}\right)^2$$

$$F = mg - f \qquad\qquad [5.69]$$

EXAMPLE 5.9.–

Let us look at the case of a resistance that opposes the movement. The force f is of the form:

$$f = K \, S \, \varrho \, v^2 \qquad\qquad [5.70]$$

with K: dimensionless coefficient (the "shape" coefficient); S: surface area; ϱ: density.

(see equations [5.68] and [5.67]) $mg \dfrac{v^2}{v_\ell^2} = K \, S \, \varrho \, v^2$

and thus:

$$K = \dfrac{mg}{S \, \varrho \, v_\ell^2} \qquad\qquad [5.71]$$

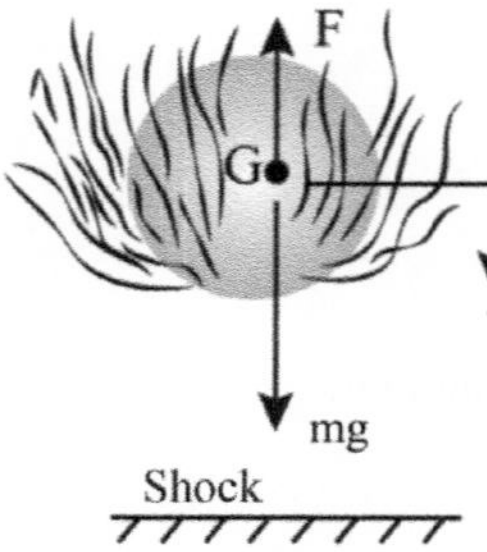

$\varrho_{air} \simeq 1.3$ g/l or 1.3 kg/m^3

$m = 30$ g, $r = 1$ cm

$K = \dfrac{mg}{\pi r^2 \varrho \, v_\ell^2}$ if $v_\ell = 40$ m s^{-1}, we have:

$K \simeq \dfrac{30 \,.10^{-3} \times 10}{\pi \; 10^{-4} \times 1{,}3 \times 1600} \simeq 0.46$

Figure 5.16. *Movement of a spherical body (ball, etc.) through the air*

Total energy change: between t and t + dt, the kinetic energy variations d are determined E_c and potential dE_p:

$$dE_c = d \left(\frac{1}{2} \, m \, v^2 \right) = m \, v \, \frac{dv}{dt}$$

with $m \dfrac{dv}{dt} = m \, \gamma = F$, and thus $dE_c = F \, v \, dt$ and with F equal to:

(see equation [5.68]) $F = mg \left(1 - \dfrac{v^2}{v_\ell^2} \right)$

we obtain:

$$dE_c = m\,g\,v\,(1 - \frac{v^2}{v_\ell^2})\,dt \tag{5.72}$$

$$dE_p = -\,mg.dz$$

$$dE_p = -\,mg\,v\,dt \tag{5.73}$$

In total, the change in energy is:

$$d\,E = dE_c + dE_p$$

$$dE = -\,mg\,\frac{v^3}{v_\ell^2}\,dt \tag{5.74}$$

Due to the air resistance, the total energy of the system decreases during the movement of the solid.

5.2.6. *Elastoplastic resistance to impact and deformation of a solid*

As an example, let us examine the case of a tubular cylinder stressed under external compression by an impulse $[I]_{t_1}^{t_2}$. This impulse generates a stress σ through the thickness e of the tube (assumed to have a thin wall), in a tube with an average radius R, we have:

$$\sigma = \frac{P\,R}{e} \text{ (P: pulse pressure)} \tag{5.75}$$

We will call P_y the elastic limit pressure of the tube (e.g. $P_y = R_e$. e/R, R_e: elastic limit of the material). The pressure P of the shock causes the radial movement x of the material and for $P > P_y$ with an acceleration $\ddot{x}$ and the forces involved per unit of surface area for a material without work hardening are:

$$P - P_y = e\,\varrho\,\ddot{x} \text{ (}\varrho\text{: material density)} \tag{5.76}$$

EXAMPLE 5.10.–

Let us examine a pressure P(t) of the form (Figure 5.17(a)):

$$P(t) = P_0 \sin^2 \omega t \tag{5.77}$$

with $0 < t < T/2$ (T: period).

NOTE.– In the case of electromagnetic pressure, we have $P_0 = B_0^2/2\mu$ valid for 1 tesla $= 0.4$ MPa ($\mu = \mu_0$) or 4 bars in the skin thickness δ in a tube made of conductive material.

Equation for the movement x of the tube wall:

(see equations [5.76] and [5.77]) $P_0 \sin^2 \omega t - P_y = e\, \varrho\, \ddot{x}$ [5.78]

That is:

$$\frac{\ddot{x}}{\ddot{x}_0} = \sin^2 \omega t - \frac{P_y}{P_0} \text{ (Figure 5.17(b))}$$ [5.79]

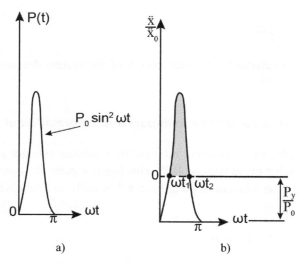

Figure 5.17. *(a) Impulse on the tube. (b) Acceleration ratio of the metal as a function of ωt and P_y/P_0*

The shaded zone represents the plastic deformation energy of the material and its kinetic energy.

In Figure 5.17(b), the "useful" launch time is $\omega\, (t_2 - t_1)$ and, by integrating t_1 and t_2, we obtain the speeds of movement $\dot{x}$, the movements x, as well as the kinetic energies of the tube $\frac{1}{2}\, m(\dot{x})^2$.

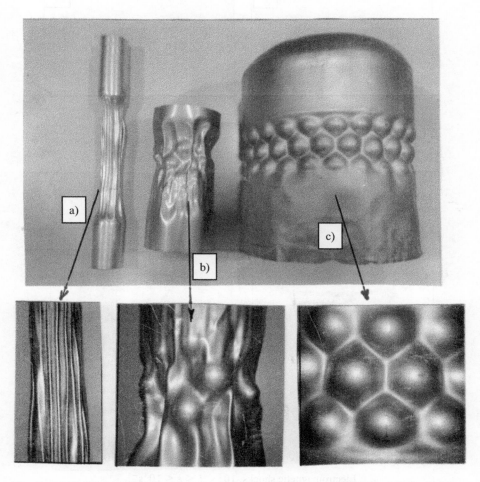

Figure 5.18. *Examples of dynamic strain of Al tubes by electromagnetic pulses. Strain speed $10^2 \leq \dot{\varepsilon} \leq 10^3 s^{-1}$; (a) an impact with dynamic buckling; (b) impact with buckling and on balls; (c) impact on balls for localized consolidation of thin-walled enclosure (propellant) (Leroy)*

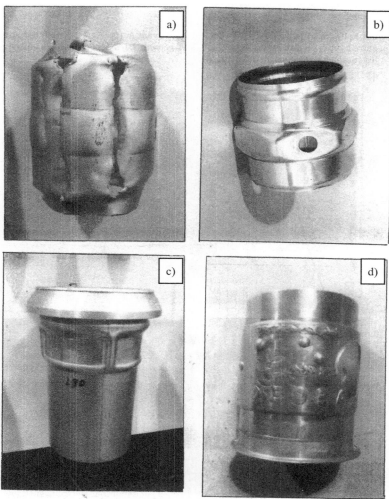

Electromagnetic shocks. $10^2 \, \text{s}^{-1} \leq \dot{\varepsilon} \leq 10^3 \text{s}^{-1}$

Figure 5.19. *Examples of shaping by expansion of tubes by the action of an internal magnetic induction; (a) free expansion and rupture; (b) a shaped part not achievable by conventional methods; (c) connection of a flange with folded down and shape in one operation; (d) shape with engraving. In (b), (c) and (d), impacts on matrices (source: Leroy, M., IUT, Nantes and Priem, D., ECN, Nantes, Magnétoformage de pièces)*

5.3. Typical loads, stress waves

Impactor geometries and impact stresses: flow diagrams of stress waves on impactors ① and forces applied over time on the receiver ②.

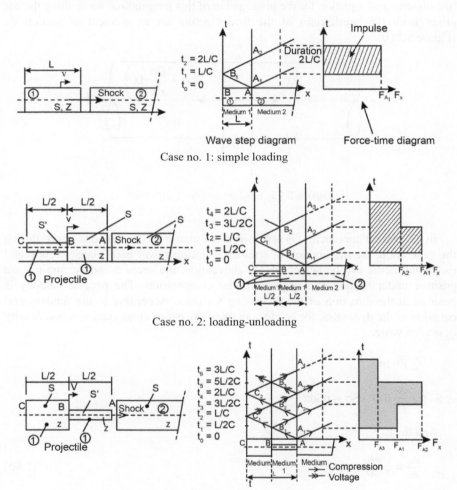

Case no. 1: simple loading

Case no. 2: loading-unloading

Case no. 3: loading-loading

Figure 5.20. *Examples of shock pulses obtained*

5.3.1. *Longitudinal compression waves, mechanical impedance, stress*

At an initial time t = 0, a longitudinal compression wave is created in a cylindrical bar by the sudden application of a force at one end (x = 0). The one-dimensional equation for the propagation of this longitudinal wave along the bar arises from the equilibrium of the forces acting on an element of section dx (Figure 5.21).

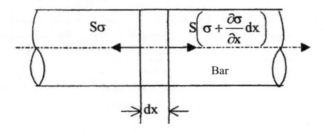

Figure 5.21. *Section element, stresses*

In the current approach, it is assumed that a planar section remains planar during the generation, the distribution of stresses is uniform over each planar section and the radial inertia can be neglected. By convention, the stress σ and the strain ε are positive under tension and negative under compression. The particle velocity is positive in the direction of the increasing x values. According to the fundamental equation of the dynamics, for an element dx of the bar of cross-section S and density ϱ, we can write:

$$\sum \vec{F} = m\vec{\gamma}$$

or $S\left(\sigma + \frac{\partial \sigma}{\partial x}\,dx\right) - S\sigma = \varrho S dx\, \frac{\partial^2 u}{\partial t^2}.$

And thus:

$$\frac{\partial \sigma}{\partial x} = \varrho\, \frac{\partial^2 u}{\partial t^2} \qquad\qquad [5.80]$$

Since the wave is elastic, the stress and the strain are connected through a linear relationship (Hooke's law) in elasticity:

$$\varepsilon = \frac{\partial u}{\partial x}$$

$$\sigma = E\, \frac{\partial u}{\partial x}$$

thus $\frac{\partial \sigma}{\partial x} = E \frac{\partial^2 u}{\partial x^2}$.

And equation [5.80] becomes:

$$\frac{\partial^2 u}{\partial t^2} - C_0^2 \frac{\partial^2 u}{\partial x^2} = 0 \tag{5.81}$$

with $C_0 = \sqrt{\frac{E}{\rho}}$ being the acoustic speed of the wave.

The general solution of equation [5.81] is written as:

$$u = f(x - C_0 t) + g(x + C_0 t) \tag{5.82}$$

It represents superimposed waves moving toward the positive values of $f(x - C_0 t)$ and toward the negative values of x $g(x + C_0 t)$.

By taking the differential of equation [5.82] with respect to x or t, we obtain the expression of the strain ε, from the stress σ and the particle velocity v:

$$\varepsilon = \frac{\partial u}{\partial x} = f'(x - C_0 t) + g'(x + C_0 t)$$

$$\sigma = E\varepsilon = E \left[f'(x - C_0 t) + g'(x + C_0 t) \right]$$

$$v = \frac{\partial u}{\partial t} = C_0 \left[-f'(x - C_0 t) + g'(x + C_0 t) \right]$$

If we consider the case of a wave propagating in a single direction, we note the existence of a linear relationship between the stress at a point and the particle velocity:

$$|\sigma| = \rho C_0 v = Zv \tag{5.83}$$

NOTE.– The product ρC_0 represents the mechanical impedance (denoted as z) that characterizes the material.

5.3.1.1. *Reflection and transmission of longitudinal waves*

When such a wave (incident wave (I)) propagating through medium 1 encounters a discontinuity or section or material, it generates a transmitted wave (T) in medium 2 and a reflected wave (R) in medium 1. The boundary conditions are as follows:

– Equilibrium of forces between the two mediums 1 and 2:

$$F_1 = F_2$$

and thus $s_1 (\sigma_I + \sigma_R) = s_2 \sigma_T$ [5.84]

– Equilibrium between particle velocities:

$$v_I - v_R = v_T$$

And thus $\dfrac{\sigma_I - \sigma_R}{z_1} = \dfrac{\sigma_T}{z_2}$ [5.85]

where s_i and z_i are, respectively, the cross-section and the impedance of the medium i (i = 1, 2); σ_I, σ_T and σ_R are the incident, transmitted and reflected stresses, respectively.

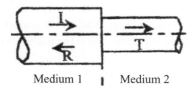

Medium 1 | Medium 2

Figure 5.22. *Stress waves*

5.3.1.1.1. Change of a wave at an interface

By combining the equations [5.84] and [5.85], we obtain:

$$\frac{\sigma_T}{\sigma_I} = \frac{2 S_1 Z_2}{S_1 Z_1 + S_2 Z_2}$$ [5.86]

$$\frac{\sigma_R}{\sigma_I} = \frac{S_2 Z_2 - S_1 Z_1}{S_1 Z_1 + S_2 Z_2}$$ [5.87]

5.3.1.2. *Reflection on a free surface*

We use the term "free surface" to mean a surface without stress. On a free surface $S_2 = 0$:

$$\sigma_R = - \sigma_I \text{ and } \sigma_T = 0 \qquad \boxed{\begin{array}{l} I \rightarrow \\ R \leftarrow \end{array}}$$

The shape of the reflected wave is the same as the incident wave, but with an opposite sign. The transmitted wave is zero, so it is completely reflected. A compression wave is therefore completely reflected in a tension wave and vice versa.

5.3.2. *Wave step diagram*

5.3.2.1. *Impact of a simple projectile*

The evolution of the wave in the projectile is studied by plotting the diagrams of the loading points and the step diagrams. A polar diagram expresses the relationship between the force and the particle velocity.

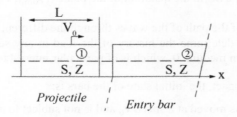

Figure 5.23. *Impact of a simple projectile*

We use the following names:

– medium 0: the entry bar of section S and impedance Z;

– medium 1: the impact bar or projectile of section S and impedance Z.

The polar diagram makes it possible to determine the state (F, v) upon the passage of the wave at the level of this discontinuity, as a function of the previous states on either side of the interface between mediums. If we call (F_i, v_i) the initial state, using equation [5.83], we obtain:

$$(F - F_i) = \pm SZ \, (v - v_i)$$

We obtain straight lines, called polar lines, with slopes $-SZ$ when the wave moves toward the positive x and $+SZ$ toward the negative x.

NOTE.– Since here we are interested in the shape of the wave that will be created in the entry bar, and which will subsequently reach the sample, we therefore only study the reflections at the projectile level only at the moment of impact (before the projectile-bar separation).

A simple case of the impact of a projectile of length L having the same section as the bars (Figure 5.23).

On the free surface of the projectile, the force is always zero. The points representative of this discontinuity will always be located on the x-axis (this axis is therefore the polar representative of the "vacuum").

In the entry bar, the impact creates a compression wave, and the particle velocity is therefore always positive (in the positive x direction) and the force is a compressive force. Just before the impact, the bar is at rest ($F = 0$ and $v = 0$), the polar representation of medium 0 is therefore a straight line passing through the origin and with a slope SZ in the quadrant of the plane (F_{comp}, v^+) (Figure 5.24).

The monitoring of the path of the waves through the different mediums over time makes it possible to determine the history of the load in any section of one of the media (in particular in medium 0, which is of particular interest to us).

Just before the impact, the initial state of the bars is:

– the impact bar is moved at a speed v_0 and is not subject to any load:

$$v = v_0 \text{ and } F = 0$$

– the entry bar is at rest: $V = 0$ and $F = 0$. The polar representation of medium 1 is a straight line passing through $(0, v_0)$ and of a slope $-SZ$ (the projectile is sent toward the positive x values):

$$F = -SZ\,(v - v_0)$$

The polar representation of medium 0 is:

$$F = SZv$$

At $t_0 = 0$, the impact (point A_1) gives rise to two compression waves moving in mediums 0 and 1. The state at this point in time at the medium-0/medium-1 interface is given by the intersection of the two poles (balance of forces), that is:

$$v_{A1} = \frac{v_0}{2} \text{ and } F_{A1} = SZ\,\frac{v_0}{2}$$

In medium 1, the wave now propagates in the negative x direction. Therefore, after impact, the polar representation of medium 1 is:

$$F = -SZ\,\frac{v_0}{2} = SZ\,(v - \frac{v_0}{2})$$

and thus $F = SZv$.

At $t_1 = It /c_0$, the wave reaches the surface (point B) and is therefore reflected integrally as a voltage wave. This wave is superimposed on the incident wave, and the result obtained by the intersection of the poles $F = SZv$ and $F = 0$ at this time gives us the state:

t_1: $F_B = 0$ and $v_B = 0$

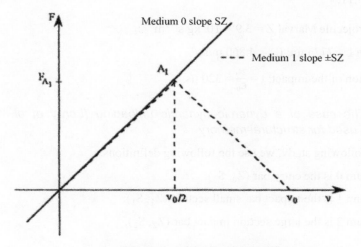

Diagram of the loading poles

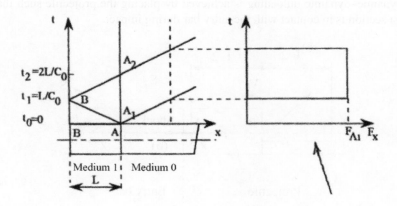

Figure 5.24. *Wave flow diagram; force-time diagram*

The wave that now propagates is a full total discharge wave, which is fully transmitted (t_2 = 2L /c_0 and point A_2) in medium 0 (with the same material and same section).

NOTE.– The length of the shock that propagates in the entry bar is l = 2L.

EXAMPLE 5.11.–

Steel projectile Marval Z = 3.9×10^7 kg s^{-1}m^{-2}:

– length L =777 mm: C_0 = 4.860 m/s;

– duration of the impact: t = $\frac{2L}{c_0}$ = 320 μs.

5.3.2.2. The case of a dynamic–dynamic unloading (Leroy et al. 1984): technique used for structural memory

In the following study, we use the following definitions:

– medium 0 is the entry bar (Z_0, S_0);

– medium 1 is the impact bar small section (Z_1, S_1);

– medium 2 is the large section impact bar (Z_2, S_2).

With Z_1 = Z_2 = Z and S_0 = S_1.

Dynamic–dynamic unloading is achieved by placing the projectile such that the largest section is in contact with the entry bar during impact.

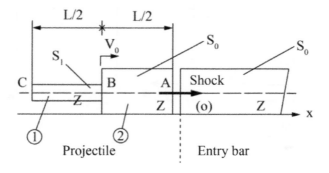

Figure 5.25. *Impact of a stepped projectile*

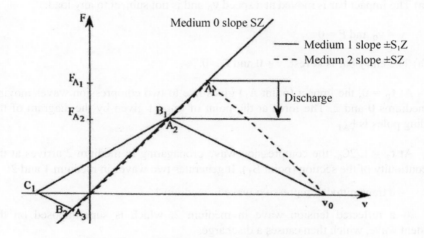

Figure 5.26. *Diagram of the unloading poles*

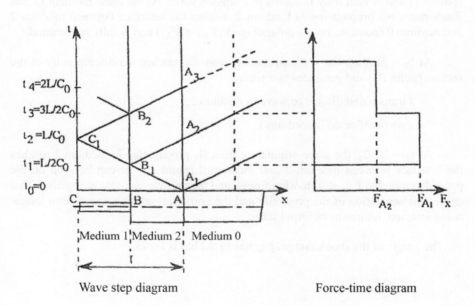

Wave step diagram   Force-time diagram

Figure 5.27. *Dynamic–dynamic discharge*

Just before the impact, the initial state of the bars is as follows:

a) The impact bar is moved at a speed v_0 and is not subject to any load:

$v = v_0$ and $F = 0$

b) The entry bar is at rest: $v = 0$ and $F = 0$.

– At $t_0 = 0$, the impact (point A_1) gives rise to two compression waves moving in mediums 0 and 2. The force at the point of impact given by the diagram of the loading poles is F_{A1}.

– At $t_1 = L/2C_0$, the compression wave propagating in medium 2 arrives at the discontinuity of the section (point B_1). It generates two waves in medium 1 and 2:

   - a transmitted compression wave in medium 1;

   - a reflected tension wave in medium 2, which is superimposed on the incident wave, which then causes a discharge.

– At $t_2 = \text{It}/C_0$, the wave propagating in the medium 1 reaches the free surface (point C_1) and is then fully reflected in a tension wave. At the same moment t_2, the discharge wave propagating in medium 2 reaches the interface between medium 2 and medium 0 (point A_2 on the polar diagram $F_{A2} < F_{A1}$) and is fully transmitted.

– At $t_3 = 3L/2C_0$, the wave emanating from C_1 reaches the discontinuity of the section (point B_2) and generates two waves:

   - a transmitted discharge wave in medium 2;

   - a wave reflected in medium 1.

– At $t_4 = 2L/C_0$, the wave originating from B_2 propagating in medium 2 reaches the interface between medium 2 and medium 0 (point A_3). It can be seen on the polar diagram that $F_{A3} < 0$ (tension force), and the particle velocity is negative. This causes the separation of the projectile and the entry bar, which since it is no longer being stressed, returns to its initial state.

The length of the shock that propagates in the bar is $l = 2L$.

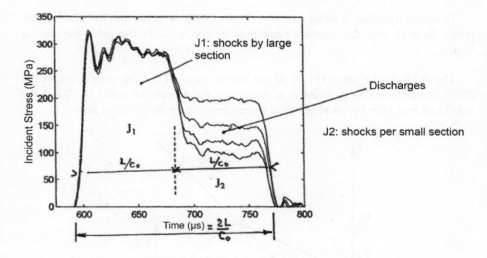

Figure 5.28. *Successive discharges by D2*

EXAMPLE 5.12.–

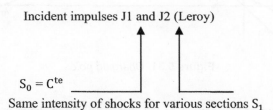

Figure 5.29. *Successive unloading shocks*

5.3.2.3. *The case of a dynamic–dynamic load*

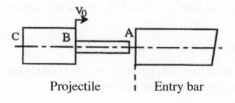

Figure 5.30. *Loading assembly*

Dynamic-dynamic loading (Leroy et al. 1984) is achieved by placing the projectile such that the smallest section is in contact with the entry bar during impact.

The study in the plane (F, V) of the intersections of the charging poles of the different media makes it possible to determine the successive states (F, V) at the interfaces, just after the passage of the wave's front, from the previous states.

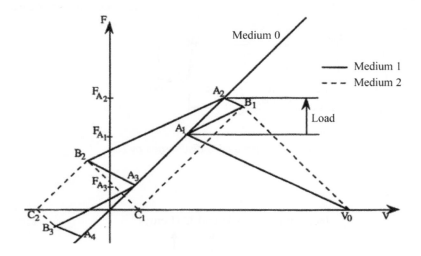

Figure 5.31. *Charging poles*

The monitoring of the course of the waves (walking diagram in the different environments) makes it possible to access the history of the loading in any section of one of the mediums.

– Just before the impact, the initial state of the bars is:

- the impact bar is moved at a speed V_0 and is not subject to any load: $V = V_0$ and $F = 0$;

- the entry bar is at rest: $V = 0$ and $F = 0$.

– At $t_0 = 0$, the impact (point A_1) gives rise to two compression waves moving in mediums 0 and 1. The force at the point of impact is F_{A1}.

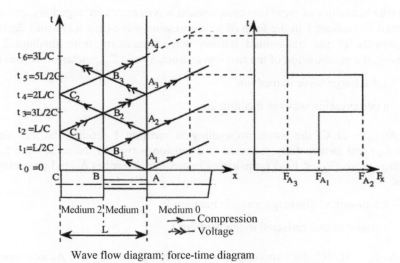

Wave flow diagram; force-time diagram

Figure 5.32. *Dynamic–dynamic load*

– At $t_1 = L/2$, the compression wave XTH "compression" propagating in medium 2 arrives at the discontinuity of the section (point B_1). It generates two waves in medium 1 and 2:

- a transmitted compression wave in medium 2;

- a reflected compression wave in medium 1.

– At $t_2 = It/$, the wave propagating in medium 1 reaches the free surface (point C_1) and is then fully reflected in a tension wave. At the same time t_2, the load wave propagating in medium 1 reaches the interface between medium 1 and medium 0 (point A_2) (Figure 5.32, $F_{A2} > F_{A1}$). It generates two waves:

- a transmitted discharge wave in medium 0;

- a reflected compression wave in medium 1.

– At $t_3 = 3L/2C$, two waves coming from C_1 and A_2 combine into B_2. This combination is broken down as follows: in Figure 5.32, as $F_{B2} > F_{B1}$, there is a decay, which means that the tension wave emanating from C_1 has a higher intensity than the compression wave emanating from A_2. This traction wave is partly transmitted into the medium 1 in the form of a voltage wave, and partly reflected into the medium 2 in the form of a compression wave. The compression wave coming from A_2 is partly transmitted in medium 2 in the form of a compression wave (the intensities of these two compression waves are added together), and partly deflected in medium 1 in the form of a compression wave (this wave thus decreases the intensity of the transmitted tension wave emanating from medium 2). In summary, the combination of the two waves from A_2 and C_1 generates two waves:

- a discharge wave in medium 1;

- a compression wave in medium 2.

– At $t_4 = 2L/C$, the wave propagating in medium 1 reaches the free surface (point C_2) and is then fully reflected in a tension wave. At the same time t_4, the discharge wave ($F_{A3} < F_{A2}$) propagating in medium 1 reaches A_3 and generates two waves:

- a transmitted discharge wave in medium 0;

- a tension wave reflected in the medium 1.

– At $t_5 = 5L/2C$, the two voltage waves coming from C_2 and A_3 combine into B_3, and generate two waves:

- a discharge wave in medium 1;

- a wave in medium 2.

– At $t_6 = 3L/C$, the wave coming from B_3 and propagating in medium 1 reaches the interface between medium 1 and medium 0 (point A_4). Figure 5.32 demonstrates that $FA_4 < 0$ (tension force), and the particle velocity is negative, thus the projectile separates from the bar.

5.4. Dynamic tests, Hopkinson technique, laws of behavior

5.4.1. *Principle of dynamic tests with Hopkinson bars*

EXAMPLE 5.13.–

– Dynamic compression and description of the test mechanisms are shown in Figure 5.33.

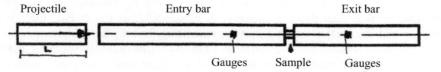

Figure 5.33. *Testing machines*

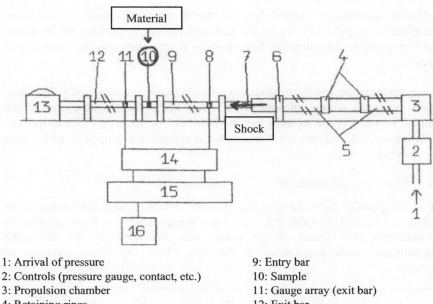

1: Arrival of pressure
2: Controls (pressure gauge, contact, etc.)
3: Propulsion chamber
4: Retaining rings
5: Cannon barrel
6: Guiding bearing
7: Projectile
8: Gauge array (entry bar)

9: Entry bar
10: Sample
11: Gauge array (exit bar)
12: Exit bar
13: Hydraulic shock absorber
14: Amplifier
15: Oscilloscope
16: PC

Figure 5.34. *Description of the test machine*

The test method consists of a projectile, an entry bar and an exit bar, all three of which have the same diameter and are made of the same material (maraging steel). The sample is positioned between the two bars.

– Generation of the stress: the impact of the projectile against the entry bar generates a stress wave compression which propagates at the speed $C_L = (E/\varrho)^{1/2}$, where E and ϱ are the modulus of elasticity and the density in the bar, respectively. The duration of the stress is the time it takes the wave to travel out and back within the projectile. When the incident compression wave reaches the interface between the entry bar and the sample, it divides into a wave reflected in tension and a wave transmitted in compression. The reflected wave returns to the entry bar, and the transmitted wave is completely restored in the exit bar.

– Useful information: knowledge of the history of elastic deformations (extensometry gauges) of the input and exit bars during the passage of the waves makes it possible to determine the history of the deformation and the stress within the sample.

– Experimental device: the projectile is propelled using a gas cannon (which uses nitrogen) operating with a solenoid valve. When firing, the kinetic energy of the impact bar is stored in the damping bar. With this device, the safety bars move very little during the test, and only the damping bar is ejected and stopped by a hydraulic shock absorber.

5.4.1.1. Choice of bar materials

It is not possible to make a choice a priori without taking into account the materials to be tested. In Table 5.2, we present a non-exhaustive list of commonly used materials. The speed of the waves was measured directly from the one-dimensional wave propagation in the bars. These values are given as an indication.

Material	E, GPa	ϱ, kg/m^3	C_L, m/s	$Z*10^6$, kg/m^{2}s
35 NCD 16	206	7,800	5,130	41.0
STUB	192	7,800	4,960	38.7
MARVAL	183	8,060	4,700	37.9
VASCO	178	8,060	4,700	36.8
2024 T3	70	2,790	5,110	14.3
PMMA	6	1,190	2,240	2.6

Table 5.2. *Average acoustic impedance of a selection of materials (Dormeval and Stelly 1980)*

5.4.1.2. Geometry of the sample

The validity of the results obtained in dynamic compression using Hopkinson bars is highly dependent on the assumption made about the homogeneity of the stresses and deformations in the sample. The first condition is a sufficiently short sample length, so that the propagation time of the waves inside it is negligible compared to the duration of the test. But other conditions, such as those that arise from the effects of inertia and friction at the interfaces, impose a well-defined geometry on the sample.

Davies and Hunter (1963) have shown that as a first approximation, the true stress in the sample can be given as:

$$\sigma = \sigma_m + \varrho \left(\frac{1}{6}\ell^2 - \frac{\ell}{8}v^2 d^2 \right) \ddot{\varepsilon}$$

with:

- σ_m: stress measured;

- ϱ: density of the sample;

- v: Poisson's ratio of the sample;

- ℓ: length of the sample;

- d: diameter of the sample.

It is noted that the inertia effects arising may be neglected if the second term is low σ_m, that is:

- if $\frac{\ell}{d} \cong \sqrt{\frac{3}{4}}\, v$, which gives d $\cong 4\ell$ for $v = 0.3$;

- if the strain speed is constant ($\ddot{\varepsilon} = 0$) but hardly achievable. It can be assumed, however, that a dynamic compression test includes two stages:

    - a very short stage (lasting a few microseconds) during which the material undergoes a very high acceleration or deceleration (during unloading) of strain,

    - a stage where the deformation speed decreases slowly, which results in a low value of $\ddot{\varepsilon}$.

If this rough scheme is respected, the effects of inertia will be important only in the phases of speed variation in rapid strain. This is what Lindholm (1964) demonstrated by carrying out tests on aluminum samples to differentiate the values of the ratio ℓ/d.

5.4.2. *Measurement of strain by extensometry gauges*

Extensometers with resistant wires are widely used at present. They are essentially constituted by a wire, generally made of constantan, from 0.015 to 0.020 mm in diameter, glued in a zig-zag pattern on a very thin support made of epoxy resin. The entry wires are connected to the ends of the resistor.

The extensometer, or resistant wire gauge, is glued to the surface of the sample to be studied, in such a way that the direction of the wires coincides with the direction of expansion which is sought to be measured. When the sample undergoes an elongation Δl in the direction of the wires, the wires undergo the same elongation, and the ohmic resistance varies. Experience shows that the relative variation of resistance and the linear elongation are connected through the expression:

$$\frac{\Delta R}{R} = k\frac{\Delta \ell}{\ell}$$

The coefficient k, known as the gauge coefficient, depends on the type of gauge, but is generally equal to about 2. The measurement of the relative variation in resistance is obtained using a Wheatstone bridge. To eliminate the influence of temperature changes, an identical gauge that is glued to a part independent of the structure and free of stresses is introduced into the measuring bridge.

Electric gauges have variable dimensions and shapes. The shortest measurement bases are about 1 mm, the longest about 50 mm.

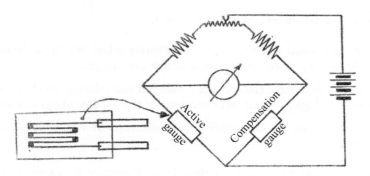

Figure 5.35. *Measuring bridge*

These make it possible to detect linear expansions in the order of 10^{-6} (some special equipement makes it possible to achieve 10^{-7}). These gauges can be used for static or dynamic measurements.

5.4.2.1. *Choice of gauges and glues*[2]

The extensometric gauges used for the measurements of deformations on the surface of the bars are selected from the classic series of manufacturers. For certain torsion applications, semiconductor gauges have been used because of their high

---

2 Excerpt from Dormeval and Stelly (1980).

gauge coefficient (K = 100, instead of 2 for conventional gauges), but their implementation, often more expensive, requires much more precautions.

The length of the grids is an important parameter. This characteristic directly influences the mechanical bandwidth of the gauge. We will try to make assemblies with gauges of the smallest possible grid length in front of the diameter of the bars. In Figure 5.36, we present a recording of strain signals obtained with grid lengths of 3.18, 1.57 and 1.14 mm, mounted on bars with a diameter of 30 mm, for a stress level of 130 MPa. Notice the reduction in amplitude (16%) and the delay at the first peak (1.5 μs) between the gauges of 1.14 and 3.18 mm.

We carried out comparative tests of the responses of gauges glued with cyanoacrylate and epoxy-type adhesives. We did not note any significant differences from the moment when the "polymerization" times and the conventional resin-hardener proportions were respected. But as the cyanoacrylate glues were more sensitive to shear, we preferred the use of epoxy adhesives.

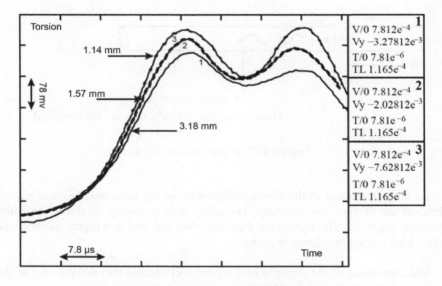

Figure 5.36. *Influence of the length of the grid on the strain response (excerpt: Dormeval and Stelly 1980)*

5.4.2.2. *Other measurement mechanisms*

The deformations obtained from the signals of the gauges implanted on the incident and transmitted bars only allow access to measurements of global deformations of the sample. If we want to access local information, we must use local observation methods:

– gauges implanted onto the test sample;

– fast cinematic observation using a slit camera (followed by a network);

– optical extensometer;

– interferometer (Michelson, Visar, Doppler).

This information makes it possible to highlight much finer phenomena, due to the heterogeneous nature of the deformations along the axis of the test piece, during the first instants of stress (a stage where the test sample is not in overall equilibrium). We can also rely on numerical simulations.

5.4.3. *Data acquisition*

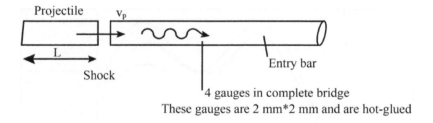

4 gauges in complete bridge
These gauges are 2 mm*2 mm and are hot-glued

Figure 5.37. *Projectile-bar shock*

As they are related to the elastic deformation of the bars, the electrical signals obtained are of very low intensity. Therefore, it is necessary to shield the entire electrical assembly. The signals are then amplified and sent to a digital oscilloscope with a high vertical resolution memory.

The conversion of the voltages read on the oscilloscope into deformation of the bars is carried out after a calibration:

– ε: strain of the bars;

– U_s: voltage read on the oscilloscope in Volts;

– G: gain of the amplifier.

The signals are then transferred and stripped on a microcomputer.

EXAMPLE 5.14.–

$$\varepsilon = \frac{2\,e_s}{V.K\,(1+v)} \text{ with } e_s = \frac{U_s}{G}$$

→ Oscillation voltage

→ Gain amp

where:

– G = 100;

– V ~ 10 V CC;

– v = 0.3;

– K ~ 2 gauge factor.

$$\varepsilon = 0.0769\ e_s = 769 \times 10^{-6}\ U_s$$

That is:

$$\varepsilon = 769\ \mu D \text{ for } U_s = 1\ V$$

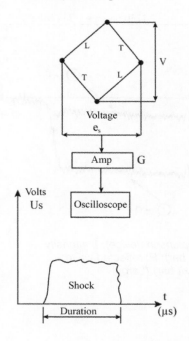

$$\frac{e_s}{V} = \frac{K}{4}\,(\varepsilon + v\varepsilon + \varepsilon + v\varepsilon)$$

$$= \frac{K\varepsilon\,(1+v)}{2}$$

$$e_s = \frac{U_s}{G}$$

with:

V: voltage supply to the bridge $\simeq$ 10 volts DC
e_s: response signal of the bridge
K: gauge factor $\simeq$ 2
ε: Deformation bar and gauge
v: Poisson's ratio of gauge
G: gain amp $\simeq$ 100
U_s: Oscilloscope signal with memory

Figure 5.38. *Acquisition of shocks*

Let us examine the case of a shock from a projectile with a speed of V_p on a bar that includes a bridge of gauges, where the projectile and bar are of the same section and material. The stress of the shock is equal to:

$$\sigma = \frac{Z}{2} V_p = E\,\varepsilon$$

For E = 190 GPa: if the oscilloscope signal = 1 V, we have σ = 146 MPa, and for bar $\emptyset$ = 20 mm, the force F of the shock is equal to:

$$F = \sigma.\,S = 146 \times 10^6 \times \pi\,(10 \times 10^{-3})^2 = 45.8 \times 10^3 N$$

Given that the duration of the shock t = 2l/c, with l as the projectile length and C as the speed of the wave, we obtain, for l = 800 mm and C = 5 × 10³m/s, a shock duration equal to: t = 320 μs, impulsion: Fxt = 14.6 N. s.

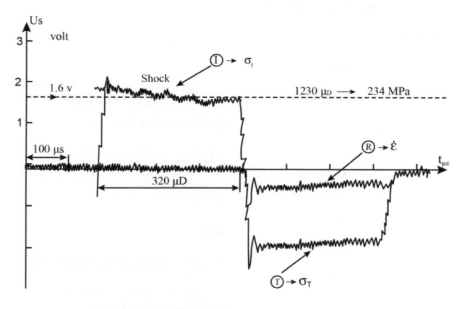

Figure 5.39. *Example of shock (Hopkinson device); I: intensity and duration of the shock (entry bar); R: reflected wave; T: transmitted wave (exit bar) (Leroy)*

5.4.4. *Data processing: analysis of dynamic behavior tests of materials*

EXAMPLE 5.15.– Case of compression in Hopkinson bars.

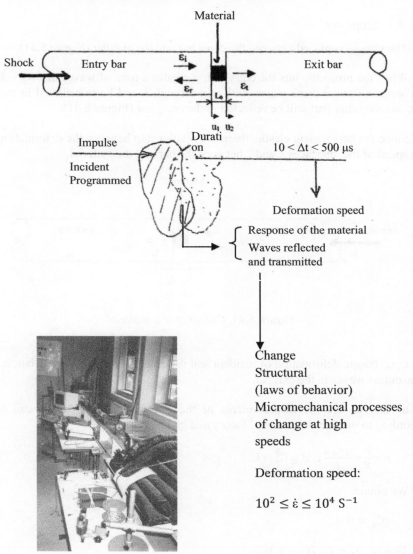

Figure 5.40. *Dynamic compression device (total length = 7m)*

For our part, the ratio l/d is equal to 1. Thus, our samples are cylinders of 8 mm in length and 8 mm in diameter. During the test, the sample/bar interfaces are lubricated so as to reduce friction.

5.4.4.1. *Stripping*

The sample is placed between the entry bar and the exit bar (Figure 5.41).

When the projectile hits the entry bar, it creates a train of waves that will charge the sample. Part of these waves will pass through it and be transmitted in the exit bar, and the other part will be reflected in the entry bar (Figure 5.41).

Since the bars remain elastic, there is a relationship between the deformation and the speed of the interfaces 1 and 2 between the bar and the material.

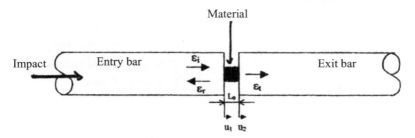

Figure 5.41. *Charging of a material*

ε_i, ε_r: elastic deformation of incident and reflected waves in the entry bar, and ε_t: transmitted waves in the exit bar.

u_1 and u_2 are the displacements of the interfaces 1 and 2. We can write, according to the one-dimensional theory and given that:

$$\varepsilon = \frac{\sigma}{E} = \frac{\varrho \, C_0 \, \dot{u}}{E} \text{ and } \varrho \, C_0^2 = E$$

We obtain:

$$\varepsilon C_0 = \dot{u}$$

That is: $\dot{u}_1 = C_0 \, (\varepsilon_r - \varepsilon_i)$:

$$\dot{u}_2 = - \, C_0 \, \varepsilon_t$$

And thus: $u_1 = C_0 \int_0^t (-\varepsilon_i + \varepsilon_r) dt$.

And: $u_2 = -C_0 \int_0^t \varepsilon_t \cdot dt$.

We thus obtain:

$$L = L_0 - (u_1 - u_2)$$

– L: length of the sample at each moment;

– L_0: initial length of the sample.

The rational deformation is given by:

$$\int_{L_0}^{L} \frac{dL}{L} = \ln \frac{L}{L_0}$$

and thus:

$$e^\varepsilon = \frac{L}{L_0} \qquad [5.88]$$

The deformation speed of the sample is:

$$\dot{\varepsilon} = \frac{1}{L} \frac{\partial L}{\partial t} = \frac{1}{L} \left(\frac{\partial u_2}{\partial t} - \frac{\partial u_1}{\partial t} \right) = \frac{C_0}{L} \left(\frac{\partial u_2}{\partial x} - \frac{\partial u_1}{\partial x} \right) \qquad [5.89]$$

By combining the equations, we obtain:

$$\int_0^\varepsilon e^\varepsilon \, d\varepsilon = - \int_0^t \frac{C_0}{L_0} \left(\varepsilon_t - (\varepsilon_i - \varepsilon_r) \right) dt \qquad [5.90]$$

The forces at the ends of the sample are given by:

$$F_1 = E_b S \left(\varepsilon_r + \varepsilon_i \right) \text{ and } F_2 = E_b S \varepsilon_t$$

where E_b is the Young's modulus of the bars and S is the section of the bars.

If we assume that the forces in the sample are in equilibrium:

$$F_1 = F_2 \text{ and thus: } \varepsilon_r + \varepsilon_i = \varepsilon_t \qquad [5.91]$$

We obtain, on the one hand, the rational strain of the sample:

$$[e^\varepsilon]_0^\varepsilon = - \frac{2C_0}{L_0} \int_0^t \varepsilon_r dt$$

and thus:

$$\varepsilon = \ln \left(1 - \frac{2C_0}{L_0} \int_0^t \varepsilon_r dt\right)$$

[5.92]

NOTE.– Therefore, the strain in the test piece is a function of the area of the reflected wave.

On the other hand, analogous to the above, we obtain the true stress in the sample:

$$\sigma = \frac{F_2}{S_{ech}} = \frac{S}{S_{ech}} E_b \varepsilon_t$$

In the plasticity of the sample, the volume is conserved:

$$S_{ech} L = S_{0ech} L_0 \Rightarrow S_{ech} = S_{0ech} \frac{L_0}{L}$$

with:

– S_{ech}: surface area of the sample at each moment;

– S_{0ech}: initial surface of the sample:

$$S_{ech} = S_{0ech} \frac{1}{e^\varepsilon}$$

and thus:

$$\sigma = \frac{S}{S_{0ech}} e^\varepsilon E_b \varepsilon_t$$

[5.93]

NOTE.– The stress in the test sample is therefore proportional to the amplitude of the transmitted wave.

We deduce from the equations:

$$\dot{\varepsilon} = - 2 \frac{C_0}{L} \varepsilon_r$$

NOTE.– The deformation speed is therefore proportional to the amplitude of the reflected wave.

The plastic deformation of the sample is obtained by removing the elastic part:

$$\varepsilon_p = \varepsilon - \frac{\sigma}{E_{ap}}$$

where E_{ap} is the apparent Young's modulus of the sample.

EXAMPLE 5.16.–

Dynamic rupture of sheath by internal shock.

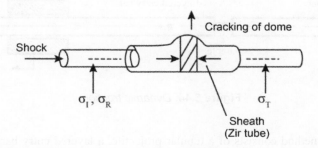

Figure 5.42. *Schematic diagram, dynamic deformation of tubes for study of internal shock resistance deformation speed > 10^2 s^{-1}*

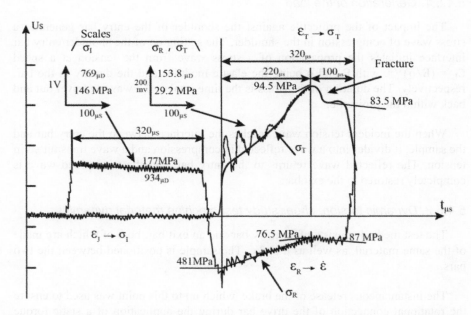

Figure 5.43. *Formation and cracking of a dome under the action of an internal compression launcher. Signal $s\sigma_I$: incident shock wave, σ_R: reflected wave and σ_T: transmitted wave. Study for resistance of sheaths to impact (20°C ≤ T < 600 °C), application to nuclear (hold of reactor sheaths under shocks) (Leroy)*

5.4.5. *Dynamic tension and torsion*

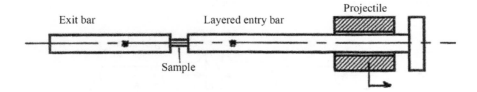

Figure 5.44. *Dynamic traction*

The test method consists of a tubular projectile, a layered entry bar and an exit bar, all three of which are made of the same material (maraging steel). The sample is screwed to the two bars.

5.4.5.1. *Generation of the load*

The impact of the projectile against the shoulder of the entry bar generates a stress wave of compression in the shoulder. The setting up of the shoulder/entry bar interface initiates the propagation of a stress wave from the tension at a speed $C_L = (E/\varrho)^{1/2}$, with E and ϱ being the elastic modulus and the density in the bar, respectively. The duration of the stress is the time it takes the wave to travel out and back within the projectile.

When the incident tension wave reaches the interface between the entry bar and the sample, it divides into a wave reflected in compression and a wave transmitted in tension. The reflected wave returns to the entry bar, and the transmitted wave is completely restored in the exit bar.

5.4.5.2. *Dynamic torsion: shear study test machine material dynamics*

The test method consists of an entry bar and an exit bar, both of which are made of the same material, as well as a brake. The sample is positioned between the two bars.

The instantaneous release of the brake, which up to this point was used to ensure the rotational connection of the drive bar during the application of a static torque over part of its length, initiates the propagation of a stress wave of torsion at the speed $C = (G/\varrho)^{1/2}$, with G and ϱ being the shear modulus and the density in the other part of the bar, respectively. The duration of the stress is the time it takes the wave to travel out and back in the pretorsion section of the entry bar.

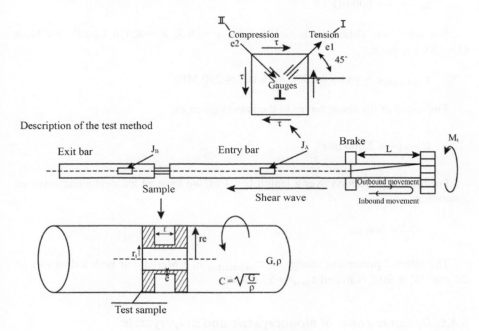

Figure 5.45. *Dynamic torsion*

When the incident torsion wave reaches the interface between the entry bar and the sample, it divides into a reflected wave and a transmitted wave. The reflected wave returns to the entry bar, and the transmitted wave is completely restored in the exit bar.

The reflected wave allows us to obtain the shear rate $\dot{\gamma}$ (so γ_t), and the transmitted wave, the stress $\tau_{(t)}$ of the tested material.

Limitation due to measurement gauges: the gauges J_A and J_B are glued to the bars and receive the elastic strains at the radii R of the bars. If the elongation e of the extensometry gauges (in security of outfit) is limited to 2,000 μD ($2{,}000 \times 10^{-6}$), we then have a maximum allowable shear of the measurement bars equal to:

$$\tau_{\text{admissible}} = \gamma\,G = (e_1 - e_2)\,\frac{E}{2\,(1+\nu)}$$

$e_1 > 0$, $e_2 = -e_1$ (gauges $\perp$ placed in primary directions I and II on the trees):

$e_1 - e_2 = 4{,}000\ \mu D$

For steel bars (Marval): $E = 190$ GPa, $\nu = 0.3$, $e = 8{,}070$ kg/m^3, we have $G = 73$ GPa, giving:

$\tau_{admissible} = \gamma\,G = 4.10^{-3} \times 73.10^9 \simeq 290$ MPa

The speed of the shear waves in the bars is given as:

$$C = \sqrt{\frac{G}{\varrho}} \simeq 3{,}000\ \text{m/s}$$

For a static pretorsion over a length $L = 1$ m, we have an incident shear wave of a duration:

$$t = \frac{2\,L}{C} = 666\ \mu s$$

The internal pretorsion energy for $\tau_{admissible}$ in a Marval bar with a diameter of 20 mm: $W_i \simeq 90.5$ N.m and $\tau_{max} \simeq 57.6$ MPa.

5.4.6. *Dynamic shear of monocrystals and polycrystals*

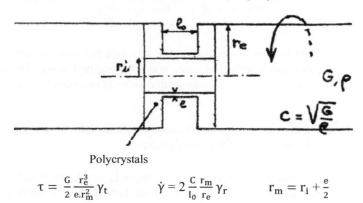

Polycrystals

$$\tau = \frac{G}{2}\frac{r_e^3}{e.r_m^2}\gamma_t \qquad \dot{\gamma} = 2\frac{C}{l_0}\frac{r_m}{r_e}\gamma_r \qquad r_m = r_i + \frac{e}{2}$$

Figure 5.46A. *Example of test samples allowing for the study of shear by dynamic torsion*

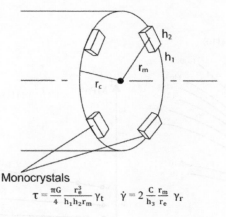

Figure 5.46B. *Example of test samples allowing for the study of shear by dynamic torsion (continuation)*

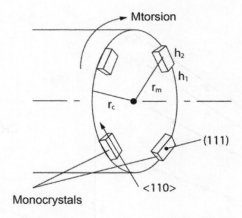

Figure 5.47. *Shearing of single crystals of aluminum stressed in the direction <110> on the plane (111)*

Figure 5.48(a) shows the change in the consolidation curves according to the deformation rate ($4 \times 10^{-5}\text{s}^{-1} < \dot\gamma < 1.6 \times 10^{3}\text{s}^{-1}$), we note that up to $\dot\gamma$ around $5 \times 10^{2}\text{s}^{-1}$, the stress increases in a linear progression by plotting (Figure 5.48(b)) the stress τ under $\dot\gamma$ (logarithmic scale). This very small increase varies with the rate of strain.

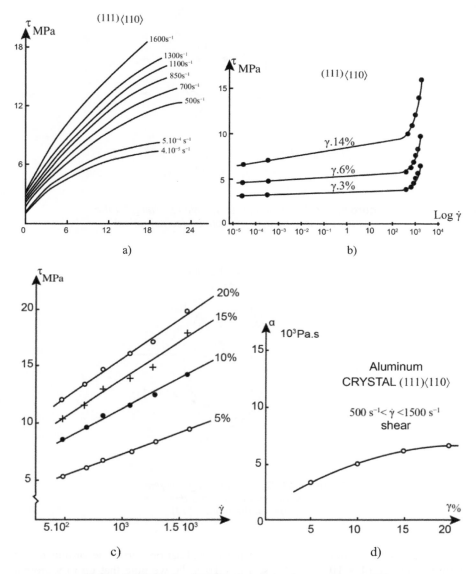

Figure 5.48. *Dynamic behavior of aluminum monocrystals under shear in <110> of the plane (111) (Leroy et al. 1979)*

The coefficient of sensitivity to the strain rate, which can be defined by:

$$m_d = \left(\frac{d\tau}{d\log\dot\gamma}\right)_T = 1/\beta$$

Equal to about 0.64 MP$_a$ for a difference in speed of six orders of magnitude and a strain of approximately 15%. For strain speeds greater than $5 \times 10^2 s^{-1}$, the stress increases more rapidly. In the linear diagram $\tau = f(\dot\gamma)$ (Figure 5.48(c)), the points are aligned and the proportionality of the stress to the deformation rate is observed.

This change in sensitivity, found by other authors, represents a change in the strain mechanism. For aluminum, polycrystalline A$_5$, the viscous coefficient α determined at the CEA DAM in compression is equal to about 3.9×10^3Pa.s. (report no. DETN/M/160/81). These results would confirm the existence of a viscous friction mechanism with:

$$\tau = \tau_0 + \alpha\,\dot\gamma$$

5.4.6.1. *Shearing of polycrystals: an example of aluminum alloy A U_4, with behavior following different types of precipitations*

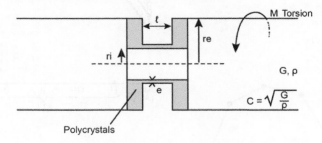

Figure 5.49. *Type of test sample*

The different structural states were obtained by the following heat treatments:

– H: placed in solution for 5 h at 530°C;

– GP: 100 h at 100°C after placement in solution;

– θ": 30 h at 160°C after placement in solution;

– θ': 10 h at 260°C after placement in solution;

– θ: 100 h at 320°C after placement in solution.

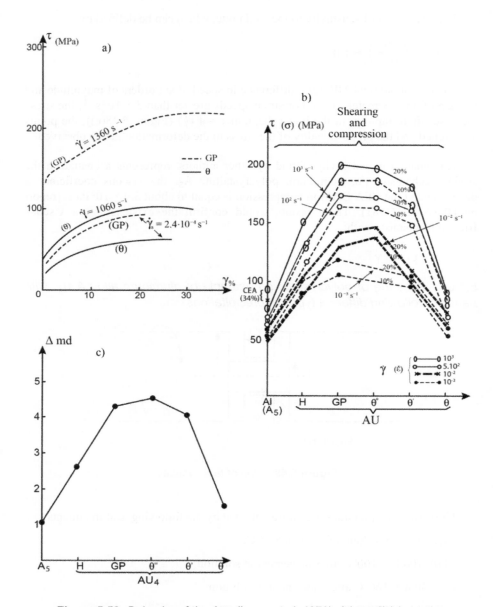

Figure 5.50. *Behavior of the AU₄ (Leroy et al. 1979). (a) τ = f(γ) in static and dynamic for θ and GP. (b) τ for the different states of precipitation. (c) Sensitivity of the stresses to the strain rates of the different states*

At quasi-static speeds, we observe a classic evolution of the stress with the state of precipitation. As in compression, from the state A_5 to θ_i, a progression in the hardening is observed for H, GP, θ'' and θ'. In dynamic, this progression is preserved but with a more marked variation, as shown, for example, by the curves $\tau = f(\gamma)$ for GP and θ (Figure 5.50(a)). The variation Δm_d of the sensitivity of the stress to the deformation rate according to the precipitation states in the aluminum matrix is deduced from the general evolution curves of τ (and σ) to isodeformation γ (and ε) for 10 and 20% for Al, H, GP, θ'', θ' and θ (Figure 5.50(b)). The conversion of values between compression and shear is deduced by adopting the ratios:

$$\gamma = \sqrt{3} \cdot \varepsilon \; \tau = \sigma/\sqrt{3}$$

Values of the sensitivities of the stresses to the strain rate of the AU4 alloy according to the different precipitation states are given as:

$$m_d = \left(\frac{d\tau}{d\log\dot{\gamma}}\right)_T$$

– Strain: $\gamma = 20\%$.

– Quasi static: $10^{-3} \; s^{-1}$.

– Dynamic: $10^3 \; s^{-1}$.

	m_d MP_a
Polycrystalline aluminum A_5	1.4
Homogenized H	3.55
GP phase	5.9
Phase θ''	6.3
Phase θ'	5.6
Phase θ	2

Table 5.3. *Values of m_d*

The change Δm_d is defined with respect to that of the aluminum matrix:

$$\Delta m_d = \frac{\left(\frac{d\tau}{d\log\dot{\gamma}}\right)_T (\text{precipitation})}{\left(\frac{d\tau}{d\log\dot{\gamma}}\right)_T (\text{Al matrix})}$$

and reported in Figure 5.50(c) for the different states of hardening at 10% and 20% of deformation in the speed range varying from 10^{-3} to 10^3 s^{-1}.

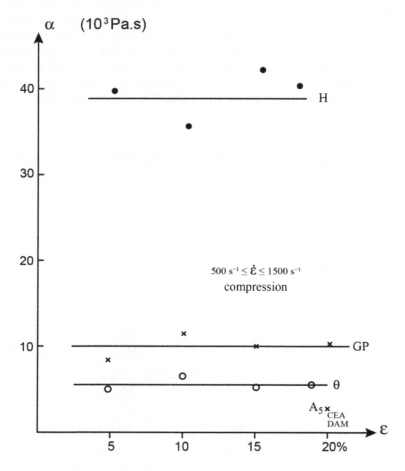

Figure 5.51. *Values of α for different deformation rates obtained between $0.5.10^3 \leq \dot{\varepsilon} \leq 15.\ 10^3 s^{-1}$, the case of H, GP and θ from the TO4 up to 20% of strain (Leroy et al. 1979)*

The similarity of the change in quasi-static curves and dynamics according to the structural states makes it possible to envision the existence of two types of displacement mechanisms of dislocations as in static.

For coherent precipitates, such as GP and θ'', crossing of the precipitates by shearing through dislocations, and for θ' and θ, semi-coherent or incoherent with the matrix, a bypass of the precipitates by the dislocations.

However, if the mechanisms are conserved, an influence is exerted on the rate of deformation represented by Δm with a variation according to the states of precipitation and, in correlation beyond $5 \times 10^2 s^{-1}$, the increase in stresses is high, with the increase resulting in a linear relationship between the stress and the strain speed.

The coefficient of internal friction B is equal to:

$$B = \rho \, b^2 \, \alpha$$

where:

– ρ: density of mobile dislocations;

– b: Burgers vector of the dislocations.

It appears that the change in α on the basis of the deformation rate is not significant when the deformation increases (Figure 5.51).

Appendix A

Primary Times of Mechanisms

This chapter will address the most important parameters that govern the strain of metals at high speeds using a dislocation model. The calculation of the principal time values of microscopic mechanisms and the influence of the viscosity of the medium on the movement of dislocations are studied within the framework of such strains.

A.1. Primary times of the behavior of dislocations

The average times necessary for crossing a dislocation, for dissociation and recombination into a perfect dislocation, for the creation and displacement of a jog, etc., are mandatory for calculating the number of dislocations created and for the movement of the dislocations causing the plastic strain.

From these primary times and for a given stress level, we can calculate the plastic strain rate $\dot{\varepsilon}_p$ as a result of the mechanisms mentioned above, as well as the variation dN of the dislocation rate (Zarka 1967).

The calculations are carried out without taking into account the parameters limiting the sliding speed, such as the electronic viscosity, the phonic viscosity, the natural inertia of the dislocation, etc. The contribution of these different parameters will be explained later.

A.1.1. *Average time to overcome a dislocation*

At the first threshold of a light slip, the movement of the dislocation is reversible; this strain is of the viscoelastic type.

When the stresses exceed the second sliding threshold, jogs form at each intersection with the other dislocations. Thermal activation helps this formation. The frequency with which this process occurs is:

$$v_{FD} = v_0 \frac{|\vec{b}|}{d'} \cos \alpha \exp - \frac{\Delta S_G}{k} (\exp - \frac{U_G^-}{kT} - \exp - \frac{U_G^+}{kT})$$

where v_0 is the Debye frequency (10^{13} Hz).

Note that $\frac{d'}{\cos \alpha}$ is the length of the dislocation element over the entirety of $|\vec{b}|$, and ΔS_G is the entropy variation of the crystal due to the displacement of the dislocation (negligible).

$$U_G^+ \simeq 2U_{FC} + (F - F_{2G})A'$$

$$U_G^- \simeq 2 \, U_{FC} - (F - F_{2G})A'$$

where A' denotes the activation area of the process of crossing the dislocation $2U_{FC}$ is the energy of formation of the jogs at the meeting of two dislocations.

It is of the order of $2 \, A\mu \, |\vec{b}|^3$. F is the force on the dislocation from the external stresses $\bar{\bar{\sigma}}$; F_{2G} is the large slip threshold.

The component of this force in any normal unit plane $\vec{n}$ containing the vector $\vec{t}$, unit tangent to the dislocation line, is given by:

$$F = |\vec{b} \, . \, \bar{\bar{\sigma}} \, . \, \vec{n} \, |$$

Therefore, the average time of crossing a dislocation is:

$$t_{FD} = \frac{1}{v_{FD}} = \frac{d'}{\cos \alpha \, v_0} \, |\vec{b}| \frac{\exp \frac{2U_{FC} + T\Delta S_G}{kT}}{2sh \frac{(F - F_{2G})A'}{kT}}$$

If the stresses are applied for a very short time, t_{FD} is infinite and the dislocations do not cross their obstacles. The activation area is generally taken to be equal to:

$$A' \simeq const. \frac{d'}{\cos \alpha} \, |\vec{b}|$$

A crossing of a dislocation is certain to occurs if:

$$(F - F_{2G}) \, A' > 2 \, U_{FC}$$

The concentration of jogs with temperature is given by:

$$C_C \simeq \exp. \frac{-U_{FC}}{kT}$$

For $b_2 \simeq b_1$, we obtain:

$$U_{FC} \simeq \frac{1}{10}\, \mu\, b_1^2\, b_2$$

And for significantly longer multiple jogs:

$$U_{FC} \simeq \frac{\mu b_1^2\, b_2}{4\,\pi\,(1-\upsilon)}\, \text{Log}\, \frac{b_1}{b_0},\ b_0 \text{ equal to } \frac{1}{2}\,b \text{ to } 2b$$

A.1.2. *Deviated recombination and sliding time*

For the deviated slip to exist, it is necessary for it to recombine into a perfect dislocation segment; the segment can then dissociate again in the plane P'.

The dissociation in imperfect creates a band of stacking faults of width C_0 of energy f per units of length.

At equilibrium, we have: $f \simeq \dfrac{\mu\, |\vec{b}|^2}{24\,\pi\,C_0}$.

Under the effect of an applied force F, the new width becomes C, with:

$$(F - F_{1G}) + 2\, f = \frac{\mu\, |\vec{b}|^2}{12\,\pi\,C}$$

These two formulas are obtained for:

$$|\vec{b_1}| = |\vec{b_2}| = |\vec{b}|\, \frac{\sqrt{3}}{3}$$

The dislocation created in the new plane P' can only remain in this plane if the segment is long enough; it is necessary for $\ell > \ell_c$ with $\ell_c \simeq \dfrac{\text{const.}}{F' - F_{1M}^\psi}$; the constant is related to the stacking fault energy in P'.

F' is the force obtained by taking the normal $\vec{n}'$ to plan P'.

The average time obtained for recombination and deflected sliding is t_{GD}^ψ, with:

$$t_{GD}^{\psi} = \frac{const.}{(F' - F_{1M}^{\psi})\, v_0\, |\vec{b}|} \exp \frac{\Delta S_{GD}}{k} \exp \frac{U_{GD}}{kT}$$

U_{GD} being the activation energy of this process:

$$U_{GD} \simeq \frac{\mu |\vec{b}|^2}{15} \, Log\, [\frac{C^{1/2}}{|\vec{b}|}]$$

since U_{GD} is positive, we have $C > |\vec{b}|$.

A.1.3. *Climb time of a dislocation*

If the stress τ is less than $\frac{\mu}{10}$, block climbs are not possible.

Thermal activation helps the gradual rise by the process proposed by Friedel: a notch must be formed; it spreads along the line and causes the rise.

The movement of a dislocation out of its sliding plane cannot be done without the displacement of material, or the creation of gaps or interstitials. The direction of the force applied to the dislocation in P' dictates the direction of the climb. Depending on whether the climb causes an increase or a decrease in volume, gaps or interstitial atoms are created.

The minimum length of the notch is $|\vec{b}|$, and the minimum displacement is $|\vec{b}|/\sin \gamma \sin \psi$, where ψ is the angle of the rising plane P' with the sliding plane P.

The average time value t_M^{ψ} necessary for the creation and the minimum displacement of the notch on the dislocation is, considering that: $\frac{|\vec{b}|^2}{\sin \gamma \sin \psi} = A'$, with:

– A': activation area of the process;

– U_{FC}: energy of the formation of the jog;

– $U_{F\ell}$, $U_{d\ell}$: energy of the formation and displacement of a vacancy;

– U_{Fi}, U_{di}: energy of the formation and displacement of an interstitial;

– F_S: chemical return force;

– ΔS_M: change in entropy of the crystal due to the climb (negligible).

$$\frac{1}{t_M^{\psi}} = v_0 \exp - \frac{\Delta S_M}{k} \exp - \frac{U_{FC} + U_{Fi} + U_{di}}{kT}$$

$$(\exp \frac{(F' - F_{1M}^{\psi}) \, |\vec{b}|^2}{kT \, (\sin \gamma \sin \psi)} - \exp - \frac{F_S \, |b|^2}{kT \sin \gamma \sin \psi})$$

The energies of the displacement and formation of spot defects for metals (in eV) are presented in Table A.1.

	Vacancies	Interstitials
$U_d + U_F$	2 to 2.5	4 to 5
U_d	0.8 to 1.2	0.05 to 0.25

Table A.1. *Vacancy and interstitial energies*

Thus, for Cu and Al, energies are presented in Table A.2.

	U_{di}	U_{dl}	U_{Fl}
Cu	$\simeq 0.2$	0.8	0.9
Al		0.35	0.8 to 1.2

Table A.2. *Cu and Al energies*

A.1.4. *Creation time of the dislocation loop*

Frank and Read have proposed an example of the creation of dislocations: a dislocation arc of length ℓ, anchored at its two ends, is flexible and has line tension.

If an external stress is exerted, the dislocation becomes curved and can develop into spirals and emit dislocation loops. This emission is limited only by the stresses exerted on the source by the loops that are emitted.

The time required to create the loop of length $2\pi L$ from the segment AB of length ℓ blocked at its ends can be evaluated as:

$$t_c \simeq \frac{3}{2} \pi \frac{W_0}{(F - F_{1G})C_t} + \frac{1}{C_t} \sqrt{\frac{W_0 \ell}{(F - F_{1G})}}$$

where:

– $W_0 \simeq \mu \, |\vec{b}|^2$ is the energy per unit length of a dislocation;

– F_{1G} is the first sliding threshold;

– $C_t = \sqrt{\dfrac{\mu}{\varrho}}$ is the transverse velocity of the elastic waves.

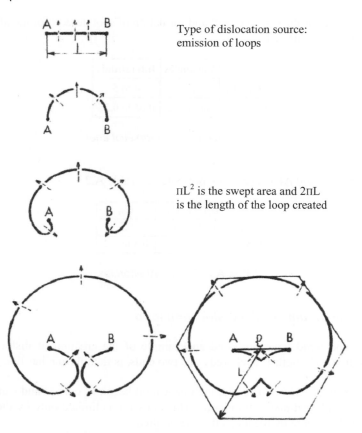

Type of dislocation source: emission of loops

πL^2 is the swept area and $2\pi L$ is the length of the loop created

Figure A.1. *Formation of a dislocation loop*

The term $\dfrac{1}{C_t}\sqrt{\dfrac{W_0\ell}{(F-F_{1G})}}$ for a short dislocation is practically negligible.

If $\ell \simeq 10^{-3}$ to 10^{-4} cm and $(F - F_{1G}) \simeq 10^{-2}\,\mu\,|\vec{b}|,\ 10^{-1}\,\mu\,|\vec{b}|$, we obtain for aluminum: $C_t \simeq 3 \times 10^5$ cm/s and $|\vec{b}| \simeq 3 \times 10^{-8}$ cm, t_c equal to 10^{-9} to 10^{-11} s.

If the stress is greater than the second threshold F_{2G} and to $F_{1G} + \dfrac{2\,W_0}{\ell}$ (the term $\dfrac{2\,W_0}{\ell}$ is obtained for a radius R of the arc equal to $\dfrac{\ell}{2}$), the equilibrium is no longer possible, and the segment is likely to be a source. A loop forms during the time t_c, to which we must add the time necessary for crossing the various dislocations that the loop has reached.

The segment can be a source if $\ell_s > \dfrac{2\,W_0}{F - F_{1G}}$, that is, for $\ell_s > \dfrac{2\,W_0}{F_{2G} - F_{1G}}$.

If the stress is low, the dislocation becomes an arc of radius R:

$$R = \frac{W_0}{F - F_{1G}}$$

The area A covered by the arc is:

$$A \simeq R\left(R_1\,\phi_0 - \frac{\ell}{2}\cos\phi_0\right)$$

$$\phi_0 = \arcsin\frac{\ell}{2R} \simeq \frac{\ell}{2R}$$

$$A \simeq \ell^3\,\frac{F - F_{1G}}{16\,W_0}$$

The number of loops emitted by each source is related to the time t_c for the creation of the loop and the time t_A moving to scan the aid $A - \pi L^2$:

$$t_A = \Sigma\,\varrho\,n'\,\pi\,L^2 \cdot t_{FD}$$

with:

– $\varrho n'$ = density per unit area of segments of a piercing system P;

– t_{FD} = average time of crossing a dislocation.

The sign Σ is extended to all piercing systems P.

During the time dt, the number of loops dP is therefore:

$$dP = K.N.\,\frac{dt}{t_c + t_A}$$

where K represents the fraction of the segments with the potential to be sources.

Given that to be a source, the segment must have a length $\ell_s > 6\,\ell$ (ℓ being the average length of segments in number N per unit mass), we obtain: $K \leq \exp - 6$.

The loop can be "blocked" by the dislocations of the piercing system P with the formation of jogs. The number of jogs on the loop is: $K_C\,(\varrho\,n'\,,\,\ell')$.

The loop breaks down into segments parallel to the main directions (e.g. 110).

The average course of a segment by deviating or climbing slip is K_{MD} (N', ℓ'); this value is an average of the shortest distances between two planes containing segments of the same system.

For systems of screw segments, the number of segments per unit of mass participating in the deflected sliding in the plane P' creating the angle ψ with P is:

$$dN'_{GD} = K_1\, N\, \frac{dt}{t^{\psi}_{GD}}$$

The number of segments per unit of mass that contribute to the rise P' is:

$$dN'_M = K_1\, N\, \frac{|\vec{b}|^2}{\ell L'\sin\gamma\sin\psi\,.\,t^{\psi}_M}$$

Given that $\dfrac{\ell L'\sin\gamma\sin\psi}{|\vec{b}|^2}\, t^{\psi}_M$ represents the time taken to cover the area L'ℓ, the segment which crosses L' is then blocked.

In the case that the stress is applied progressively, the authors show that K and K_1 are stresses.

Some of the dislocations no longer intervene in the process; it becomes possible for a segment that has moved from L to encounter a parallel dislocation of the opposite Burger vector, or for the segment to participate in the formation of a sub-joint. This fraction is assumed to be constant.

A.2. Calculation of the plastic strain rate $\frac{d\varepsilon_p}{dt}$ and the variation in the rate of dislocation segments dN_s for a polycrystal

In the case of a polycrystal (aggregate of monocrystals of any shape and extremely disoriented relative to each other), it is necessary to write the equations from Chapter 1 (section 1.3.2.3) at all points of each crystal, taking into account the continuity of the normal force and the temperature at the joints.

The problem is complex, and some approximations must be made. Taylor (1934), and Hill and Bishop (1967) connected the rate of plastic strain $\dot{\varepsilon}_p$ in a polycrystal at the sliding speed γ on each mechanism by:

$$\dot{\varepsilon}_p = \frac{1}{m}\,\Sigma\,\gamma \quad \text{(extended to all the slips of the crystals)}$$

The coefficient m makes it possible to define the split on the slip system from the applied stress σ on the polycrystal.

$$\tau \simeq \frac{\sigma}{m}$$

with:

– τ = reduced split (force per unit length of dislocation divided by the modulus of the Burger vectors);

– m = a purely geometric coefficient associated with the systems of the structure and characterizing the distribution of the crystals in the polycrystal; for CFCs m ≃ 3.06.

The force on the dislocation is $F = |\vec{b}|\frac{\sigma}{m}$.

If we use N_s to represent the number per unit mass of small segments of dislocations of a length ℓ in the polycrystal, we assume, as did Friedel (1964), that the average length of the small segments is proportional to $(\varrho\, N_s)^{-1/3}$, where ϱ is the specific mass, and that the same applies to the shortest average distance d between two parallel segments or two parallel planes containing segments.

The study of the monocrystal leads us to propose for the polycrystal (neglecting the Peierls–Nabarro forces, and assuming that the modules of the Burger vectors are all equal):

$$\tau_1 = \frac{\mu\,|\vec{b}|}{k_1\,m}\, N_\ell^{\frac{1}{2}} = \frac{\mu\,|\vec{b}|\,(\varrho\,N_s)^{1/3}}{k_1\,m}$$

and:

$$\tau_2 = \frac{\mu\,|\vec{b}|}{k_2\,m}\, N_\ell^{\frac{1}{2}} = \frac{\mu\,|\vec{b}|\,(\varrho\,N_s)^{1/3}}{k_2\,m}$$

where N_ℓ and N_s are the densities of dislocation lines per unit area and the equivalent number of segments per unit mass, m is the coefficient defined previously and K_1 and K_2 are two constants representing the annealing conditions of the metal.

The first threshold τ_1 represents a slight displacement of a dislocation in its plane or out of its plane. In this displacement, less than the average distance between the dislocations, the dislocation must then only surpass the remote actions of its own system. The second threshold τ_2 represents a significant displacement where the dislocation, in addition to surpassing the actions of its own system, must surpass the remote and contact actions of the other dislocations it encounters.

For a cubic metal with centered faces, a calculation assuming that the dislocation density is the same in all systems gives:

$$K_1 \simeq 2.2 \, K_2$$

The contribution to the slip for a density $\varrho \, N$ dislocations per unit of volume each scanning the area A in parallel displacement planes between two planes that are distant from the unit in the direction $\vec{b}$:

$$\vec{\Gamma} = \varrho \, N \, A \, \vec{b}$$

Returning to the hypothesis established by Zarka (1967), the displacements of dislocations causing an increase in plastic strains are, during the time dt:

$$\gamma \, dt = |\vec{b}| \, \Sigma \, (\varrho \, dP \, \pi \, L^2)$$

$$\gamma' \, dt = |\vec{b}| \, \Sigma \, (\varrho \, dp' \, L' \ell)$$

A.2.1. *Extended sums for all systems*

$- \gamma \, dt$ is the displacement between two planes P, which are distant from the unit in the direction $\vec{b}$.

$- \gamma' \, dt$ is the displacement between two planes P', which are distant from the unit in the direction $\vec{b}$.

The planes P' are parallel to the dislocations of the system.

dP is the number of loops (or lines) of dislocations per unit mass (the loops are broken down into segments of different systems by their orientation) created by the sources.

The loops dP cover an average area equal to πL^2 before being blocked on the dislocations which penetrate their plane P. From among the segments blocked at the time t, some of them, of a number dp' per unit mass, can move by deflected sliding or by climbing in other planes P', each over an average distance L'. Some of them, totaling to g' dp', are able to arrange themselves in sub-joints, or disappear with other parallel segments and opposite Burger vectors.

The quantities characterizing a system, (N_s, ℓ), are modified, and each loop created is broken down into segments of length $S = sL$ and, during the time dt, and we obtain:

$$dN = dP - \Sigma\, g'\, dp'$$

$$d(N_s, \ell) = dP\, sL - \Sigma\, g'\, dp'\, \ell$$

$$d\ell = \frac{dP}{N_s}\, (s\, L - \ell)$$

During the time dt, the plastic strain speed $\dot{\varepsilon}_p$ and the rate of the dislocation segments $\dot{N}_s$ is:

$$\dot{\varepsilon}_p = \frac{|\vec{b}|}{m}\, (\varrho\, \frac{dP}{dt}\, \pi\, L^2 + \varrho\, \frac{dp'}{dt}\, L'\, \ell)$$

$$\dot{N}_s = 3\, \pi\, \frac{L}{\ell}\, \frac{dP}{dt} - \frac{3}{2}\, g'\, \frac{dp'}{dt}$$

We assume that the numbers dP and dp' are of the form:

$$dP = N_\ell^{3/2}.T.dt$$

$$dp' = N_\ell^{3/2}.T'.\, dt$$

where T and T' are the inverse of the time required for a dislocation to form and sweep the area πL^2, and the inverse of the time required for a dislocation to move by deflected sliding and travel through the new plane L'.

For the creation of the dislocation, we see a simultaneous branching out of the line, which requires the time t_c, and the surpassing of obstacles, which requires the time t_F.

Similarly, for the deviated slip, we have both a combination of the dislocation into a perfect dislocation in the initial plane, which requires the time t_G, and the crossing of a critical distance in the new plane, which requires the time t_F.

The values T and T' can be taken as equal to:

$$T = \frac{1}{t_c + t_F} \text{ and } T' = \frac{1}{t_G + t_{F'}}$$

The calculations of t_c, t_F, t_G and $t_{F'}$ have been carried out previously.

If we define γ as the unit step function, meaning that if $\tau < \tau_2 \rightarrow T = 0$: $\beta\tau' < \tau'_2$ $\rightarrow$ T' = 0 (β is a coefficient defining the average stress applied on the deflected sliding plane), we can calculate $\dot{\varepsilon}_p$ and $\dot{N}_s$.

A.3. Influence of viscosity η on the behavior of dislocations, application to the calculation of plastic strain ε_p

For the stresses σ above the threshold σ_2, the dislocations cross the boundaries of the dislocation forest.

The speed of movement γ for a density N_M of dislocation segments is connected to the speed v of the segment in the form: $\gamma = N_M$ bv.

Any force F exerted on the dislocation by application of a stress σ creates a split τ in the slip plane. In the case of a polycrystal, we have:

$$F = |\vec{b}| \frac{\sigma}{m} = |\vec{b}| \tau$$

The damping coefficient B (a function of the viscosity η of the medium) relates the force F to the velocity v acquired by the dislocation as given by:

$$F = B \, v = \tau \, |\vec{b}|$$

The primary times for dislocation displacements therefore depend on η.

In the strain of metals at high speed ($\dot{\varepsilon} > 10^3$ s^{-1}), it can be considered that the strain depends mainly on the viscous damping B of the dislocations. Effectively, the length of the route d_{90} for which the dislocation reaches 90% of its limit speed v_ℓ is equal to:

$$d_{90} = 1.4 \, v_\ell \frac{m}{B}, \text{ with } v_\ell = \frac{\tau b}{B}$$

where m is the mass of the dislocation per unit length and τ is the stress exerted.

The measurements of internal friction (Alers and Thompson 1961) and ultrasonic attenuation (Hikata and Elbaum 1967; Mason 1968) cause B to be given a value of approximately 10^{-4} dyn.s.cm^{-2} at ambient temperature. If we assume that $V/10 < v_\ell < 2 \, V/3$, where V is the speed of sound, we calculate that:

$$8 \, |\vec{b}| < d_{90} < 50 \, |\vec{b}|$$

Thus, the dislocation reaches its limit speed at the end of a path measuring the length of a few atoms.

Considering that the speed of dislocations is v_ℓ in the case of large slips, we can write that the speed of plastic strain $\dot{\varepsilon}_p$ is a direct function of B and the density N_M of mobile dislocations:

$$\dot{\varepsilon}_p = \frac{N_M\, b^2\, \tau}{B}$$

The coefficient B changes according to the type of dislocation; we will use B_V and B_C, respectively, to represent the damping coefficients of a screw dislocation and an edge dislocation.

The viscosity η is a function of the temperature and depends on the electronic and phonic behavior of the medium. We will show the importance of η_e and from η_p for certain temperature domains, and we will attempt to explain the experimental strain curves of monocrystals and polycrystals of CFC and HC metals on the basis of the temperature. Now, we will calculate the coefficients B_V and B_C for screw and edge dislocations.

A.3.1. *Calculation of B_V*

The relationship between the applied shear force T_{12}, the shear stress S_{12} and its speed $\dot{S}_{12}$ can be given as:

$$T_{12} = \mu_0\, S_{12} + \eta\, \dot{S}_{12}$$

given that $\mu = \mu_0 + I\omega\eta$.

The energy loss per element of volume is:

$$(T_{12}\, \dot{S}_{12})\, dx\, dy\, dz = (\mu_0\, S_{12}\, \dot{S}_{12} + \eta\, \dot{S}_{12})\, dx\, dy\, dz$$

In the case of a screw dislocation, a cylindrical coordinate system is taken, and the shear stress at a point remote from r from the core of the dislocation is:

$$S_{\theta z} = \frac{b}{2\,\pi\, r}$$

where $\vec{b}$ is the Burgers vector of the dislocation.

Now we will consider an element of fixed volume in space. If the dislocation moves at the speed v, the value of r_1 (the distance to a point P at the angle θ) changes to r_2:

$$r_2 : (r_1^2 + (vdt)^2 - 2\, r_1\, (vdt)\cos\theta)^{\frac{1}{2}}$$

$$r_2 = r_1\, (1 - \frac{vdt}{r_1}\cos\theta)$$

The variation of the shear stress is then:

$$\Delta S_{\theta z} = \frac{b}{2\,\pi}\, (\frac{1}{r_2} - \frac{1}{r_1})$$

$$\frac{\Delta S_{\theta z}}{\Delta t} = \dot{S}_{\theta z} = \frac{\frac{b}{2\,\pi}\, (\frac{1}{r_2} - \frac{1}{r_1})}{dt}$$

$$\dot{S}_{\theta z} = \frac{b\, v\cos\theta}{2\,\pi\, r_1^2}$$

The dissipated energy is: $T_{\theta z}.\dot{S}_{\theta z} = \eta\, \dot{s}_{\theta z}^2$.

By integrating on a cylinder of unit length surrounding the dislocation:

$$W = \frac{b^2\, v^2\, \eta}{4\,\pi^2} \int_0^\infty rdr \int_0^{2\pi} \frac{\cos^2\theta}{r^4}\, d\theta = \frac{b^2\, v^2\, \eta}{8\,\pi\, a_0^2}$$

where a_0 is the radius of the center of the dislocation.

Thus, $F = T_{13}$, $b = vB$ and the dissipated energy $W = Fv$, and then:

$$W = v^2\, B = \frac{b^2\, v^2\, \eta}{8\,\pi\, a_0^2}$$

$$B_v = \frac{b^2\, \eta}{8\,\pi\, a_0^2}$$

The value of the radius of the center of the dislocation is the subject of debate (Cottrell 1953), and generally we use:

$$a_0 = \frac{\mu\, b^2}{8\,\pi^2\, \gamma}$$

where γ is a term for energy per unit area.

A reasonable value of γ is $\gamma = 1,000$ ergs/cm^2, which corresponds to $a_0 \simeq \frac{b}{6}$.

For a screw dislocation:

$$W = \frac{36 \eta v^2}{8 \pi} = 1.43 \, \eta \, v^2$$

that is, $B_v \simeq 1.43 \, \eta$ for $a_0 = \frac{b}{6}$.

A.3.2. *Calculation of B_c*

A calculation similar to the previous one makes it possible to determine the value of the viscosity B_c attributed to a wedge dislocation.

The center of the edge dislocation has a radius of:

$$a_{0c} = a_{0v} (1 - v)$$

where v is Poisson's ratio.

In this case, the calculations done by Cottrell (1953) show that:

$$S_{r\theta} = \frac{b \cos \theta}{2 \pi r (1 - v)}$$

$$\Delta S_{r\theta} = S_{rr} + S_{\theta\theta} + S_{zz} = \frac{\mu \, b \sin \theta}{3 \pi k (1 - v)}$$

with: $K = \lambda + 2 \, \mu \, / \, 3$.

The dislocation undergoes a loss of energy due to the shear viscosity η and a loss due to the compression viscosity χ. These are evaluated as before, and we find:

$$W = \frac{3}{4} \left(\frac{b^2 \, \eta \, v^2}{8 \pi (1 - v)^2 \, a_0^2} \right) + \frac{1}{72 \pi} \left(\frac{\mu^2 v^2 \chi}{K^2 (1 - v)^2 \, a_0^2} \right)$$

$$B_c = \frac{3}{4} \left(\frac{b^2 \, \eta}{8 \pi (1 - v)^2 \, a_0^2} \right) + \frac{1}{72 \pi} \left(\frac{\mu^2 \chi}{K^2 (1 - v)^2 \, a_0^2} \right)$$

$$B_c \simeq \frac{3}{4} \left(\frac{b^2 \, \eta}{8 \pi (1 - v)^2 \, a_{0v}^2} \right)$$

The values of B_v and B_c therefore depend on the determination of η and a_0.

The value of B depends on the center of the dislocation a_0. If the dislocations move relatively slowly (at a speed much lower than the speed of sound), a_0 can be determined by the nonlinear nature of the elasticity constants. In this case, Mason (1955) estimates that a_0 is in the order of 3 b/4.

The speed v depends on the stress σ and, at high stresses, this speed tends toward a limit value that is a function of B.

Johnston and Gilman (1958) were able to measure at ambient temperature the speed of the dislocation as a function of the stress applied by displacement of the corrosion figures (see Figure A.2) on lithium fluoride.

These curves show that for values beyond 2×10^8 dynes/cm^2, the rate of dislocations for lithium fluoride approaches a constant. Past 2×10^8 dynes/cm^2, the dislocations move about freely in the dislocation forest.

We have calculated the damping terms B_c and B_v as a function of η for the various dislocations. The viscosity η depends on the interactions between shear waves and particles such as electrons, phonons, etc.

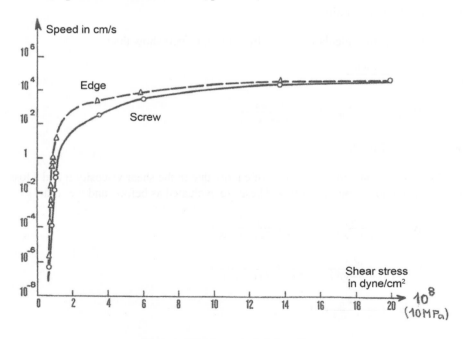

Figure A.2. *Speed of dislocations*

Below, we will make clear the terms η_p (viscosity due to phonons), η_D (broadcast of phonons) and η_e (viscosity due to electrons).

A.3.2.1. *Viscosity due to phonons*

A shear wave propagating in a metal undergoes attenuation due to a thermal mechanism called phonon viscosity.

The first stage of relaxation was studied by Mason (1964).

A.3.3. *Viscosity coefficient*

Mason showed that the sudden application of stress increases the modulus of elasticity by a quantity:

$$\Delta c = 3 \, E_0 \, \frac{\Sigma_i \, (\gamma_i^i)^2}{n}$$

where:

– E_0 = the thermal energy density;

– γ_i^i = the Grüneisen number for the mode i and to the constraint i;

– n = the number of modes.

It was Maxwell who first showed that an increase Δc of the elastic modulus, following the application of a stress, is equivalent to the viscosity η of a gas, because of the relaxation time necessary for the molecules to return to their equilibrium state.

For a solid body of a volume V_0, if V_t and V_ℓ are the transverse and longitudinal speeds of sound, the total number of vibrations in a frequency interval df is given by:

$$Z(f) = 4 \, \pi \, V_0 \, \left[\frac{2}{v_t^3} + \frac{1}{v_\ell^3}\right] f^2 \, df$$

The Debye limit frequency fd common to the two modes is:

$$f_d^3 = \frac{9 \, N}{4 \, \pi \, V_0} \left[\frac{2}{v_t^3} + \frac{1}{v_\ell^3}\right]$$

where N is the total number of frequencies equal to the number of particles in the volume. Each vibration has an energy associated with this frequency equal to:

$$\frac{\hbar w}{e^{\hbar w/kT} - 1}$$

Adding these energies together allows us to obtain E_0.

The shapes of the spectra are complex and obtaining the average Debye speed $\overline{V}$ for the various modes is difficult.

The time of the relationship τ is related to the thermal conductivity K and to the specific heat C_V by:

$$\tau = \frac{\overline{\ell}}{\overline{V}} \text{ with } \overline{\ell} = \frac{3K}{C_v \overline{V}}$$

where $\overline{\ell}$ is the average free path of the phonons.

Here, $\tau = \frac{3K}{C_v \overline{V}^2}$, and the viscosity of the equivalent gas is given as:

$$\eta = \Delta c \cdot \tau = \frac{DE_0 K_p}{\varrho C_v \overline{V}^2}$$

where:

– D = a nonlinear constant determined from the Grüneisen numbers;

– E_0 = the total thermal energy, the sum of the energies of the possible modes of vibration of the metal;

– ϱ = the density;

– τ = relaxation time between hot and cold phonons.

For a metal, the thermal conductivity occurs mainly due to the electrons, and consequently, the viscosity due to the phonons is much lower than for quartz or for semiconductors such as germanium.

For a metal, the phonic viscosity is given as:

$$\eta = \frac{E_0 K_p}{C_{vp} \overline{V}^2}$$

where K_p and C_{vP} are, respectively, the thermal conductivity and the specific heat due to the phonons.

In a practical sense, the specific heat due to the electrons is given as:

$$C_{ve} = \beta\tau$$

where β depends on the metal ($\beta_{Al} = 3.3 \times 10^{-4}$ cal per mole per degree Kelvin and $\beta_{Cu} = 1.78 \times 10^{-4}$ cal per mole).

The ratio between the thermal conductivity from electrons and that from phonons is:

$$\frac{K_e}{K_p} = \frac{C_e \, \bar{c} \, \bar{\ell}_e}{C_p \, \bar{V} \, \bar{\ell}_p} \quad \text{or} \quad \frac{K_e}{K_p} = \frac{C_{ve} \, \bar{c} \, \tau_{eP}}{C_{vP} \, \bar{V}^2 \, \tau_{Pe}}$$

$\bar{c}$ is the Fermi velocity of the electrons τ_{eP}, their relaxation time (electron–phonon interaction) τ_{Pe} is the phonon-electron relaxation time, and C_e and C_p are the specific electronic and sound heat.

The Fermi velocity of the electrons is given as a function of their number N by cm^3 and their mass m by:

$$\bar{c} = \frac{(3\,\pi^2 N)^{1/3}\,\hbar}{m}$$

Using a simple model for cubic crystals, Leibfried and Schlöman (1954) estimated K_P :

– $K_P = 3.6$ a A θ^3/γ^2 T;

– a = the length of the network;

– A constant = 92.9;

– θ = the Debye temperature;

– γ = the Grüneisen constant.

But this relationship is valid only for $T > \theta$, and it gives very incorrect results for the change of K_P with the temperature when $T < \theta/10$.

In the relationship $\frac{K_e}{K_p}$, the relaxation times intervene τ_{eP} and τ_{Pe}; according to Kittel (1956), for a metal of good purity, the ratio of these times gives the value of:

$$\frac{\tau_{eP}}{\tau_{Pe}} \simeq 10^{-2}$$

The electronic specific heat C_e for aluminum is 1.46×10^3. T ergs/cm per K, $\bar{c} = 1.43 \times 10^8$ cm/s, the average speed of Debye is $\bar{V} = 3.7 \times 10^5$ cm/s (Anderson 1956). By replacing K_P in the formula giving the viscosity η, we obtain:

$$\eta = \frac{E_0 \, K_P}{C_{vP} \, \bar{V}^2} = \frac{E_0}{C_{vP} \, \bar{V}^2} \times \frac{K_e \, C_{vP} \, \bar{V}^2 \, \tau_{Pe}}{C_{ve} \, \bar{c}^2 \, \tau_{eP}}$$

That is:

$$\eta = \frac{E_0 \, K_e \, \tau_{Pe}}{\beta \, T \, \bar{c}^2 \, \tau_{eP}}$$

Or:

$$\eta = \text{const.} \, \frac{E_0 \, K_e}{\bar{c}^2 \, T}$$

for aluminum $\frac{E_0}{T} = 3.62$ cal. per mole:

$$K_e = 2.22 \times 10^7 \text{ ergs cm}^2/\text{cm/K}$$

$$\bar{c} = 1.4 \times 10^8 \text{ cm/s}$$

for copper $\frac{E_0}{T} = 3.94$ cal. per mole:

$$K_e = 3.93 \times 10^7 \text{ ergs cm}^2/\text{cm/K}$$

$$\bar{c} = 1.58 \times 10^8 \text{ cm/s}$$

The diffusion coefficient D.

As Mason indicates, it is given by:

$$D = [3 \, (3 \, \Sigma_i \, E_i \, (\gamma_i^i)^2 - \gamma^2 \, \varrho \, C_v \, T_0] / \, E_0$$

with:

– E_0 = total thermal energy density at temperature T_0;

– E_i = thermal energy of mode i;

– γ_i^i = the Grüneisen number for the mode i and to the constraint S_i.

Thus, the constant D is evaluated for silicon and germanium by using the 39 wave modes propagating along the directions $\langle 100 \rangle$, $\langle 110 \rangle$ and $\langle 111 \rangle$. The Grüneisen numbers are calculated from the moduli of elasticity of the second order and from the measurement of the moduli of elasticity of the third order.

$$- \gamma_i^i = u_i\, u_k + \frac{N_p\, N_q}{2\, c} \left(c_{ikpq} + u_r\, u_s\, c_{ikpqrs} \right)$$

where:

- N_p, N_q = the principal cosines of the direction of propagation;

- u_i, u_k = directing cosines of the particle displacements;

- c_{ikpq} = the elastic modulus in tensor notation;

- c_{ikpqrs} = the third order elastic modulus;

- C = the elastic modulus controlling the propagation of the wave in the direction determined by N_p, N_q and by the particle movements u_p, u_s.

For example, for a longitudinal wave propagating along $\langle 100 \rangle$ with an applied constraint S_{II}:

$$- \gamma_i^i = 1 + \frac{c_{IIII} + c_{IIIII}}{2\, c_{II}}$$

We note that:

$$- \gamma_i^i = \frac{3\, c_{II} + c_{III}}{2\, c_{II}}$$

A.3.4. *Phonon diffusions*

Leibfried's theory (1950) considers two sources of diffusion:

- the diffusion of thermal phonons by the stress field of a moving dislocation;

- diffusion by the vibrations of the dislocation.

The first process leads to a braking stress in the order of:

$$\frac{E_0}{60} \times \frac{v}{c}$$

with:

- E_0 = the thermal energy density;

– V = the dislocation speed;

– c = the speed of sound.

The second leads to a stress in the order of: $\frac{E_0}{10} \times \frac{V}{c}$.

In the calculation, it is assumed that the isotropic flow of phonons incident on the dislocation is not modified by the phonon–phonon diffusions.

As a result of this work, Lothe (1960) evaluated the coefficient of damping t as:

$B_D = a\ E_0/10\ V_s$

where:

– a = the network parameter;

– V_s = the speed of the shear wave.

For aluminum, a = 4.05×10^{-8} cm, $V_s = 3.03 \times 10^5$ cm/s.

$E_{0_{300K}} = 44.25 \times 10^8$ ergs/cm^3

B_D is usually low with regard to B_p; for aluminum at ambient temperature, we have:

$$\frac{B_D}{B_{P_{300K}}} \simeq 4.5 \times 10^{-2}$$

A.3.5. *Influence of electric waves*

Mason described the electron–dislocation interactions by a mechanism of the same type as that studied for the phonon–dislocation interactions.

Pippard showed that the electric and magnetic fields produced by a stress wave in a metal will force the electrons to move in such a way as to maintain a neutral current. The result is that the electrons provide a speed equal to the speed of the particle of the wave. The moment allotted to the electrons can be exchanged between surfaces moving at slightly different speeds, and it produces an electronic viscosity that can dampen a wave propagating in the metal.

Mason, Pippard and Morse calculated the viscosity coefficient for a spherical Fermi surface:

$$\eta_e = 9 \times 10^{11} \, \hbar^2 \, (3 \, \pi^2 \, N)^{2/3}/5 \, e^2 \, \varrho$$

with:

– $\hbar = h/2 \, \pi$;

– e = electronic charge;

– N = the number of electrons per cm^3;

– ϱ = the resistance.

The damping coefficient can be related to the viscosity by applying the formulas previously established for screw and wedge dislocations. The value of B_e still depends on the radius of the center of the dislocation. Blount (1959) showed that once the stress becomes very large near the dislocation, the damping becomes saturated.

The condition for saturation established by Blount is:

$$q \, \ell_e \, S \, E/m \, v^2 \simeq 1$$

where:

– q = number of orders;

– ℓ_e, m, v = average free path, mass and velocity of the electron.

In the case of a potential strain E in the order of 5 eV (generally associated with a spherical Fermi surface), this represents a stress S of 5×10^{-2}, equivalent to a radius a_0 of 10^{-7} cm.

It is generally accepted that the center a_0 of the dislocation is equal to 10^{-7} cm, which gives an electronic damping coefficient B_e equal to:

$$B_e = 3.25 \times 10^{-3} \, \eta_e$$

A.4. Supplemental materials

Appendices B, C and D are available for download at: www.iste.co.uk/leroy/rheology1.zip:

– Appendix B: Calculation of Viscous Damping Coefficients on the Basis of Temperature for: Aluminum, Lead, Copper, Gold, Silver, Zinc;

– Appendix C: Resistivity, Energy Density, Thermal Conductivity and Specific Heat;

– Appendix D: Excerpts: Publications on Dynamic Rheology.

References

Chapters 1–5

Alers, M. (1970). *Metal Trans.*, 1, 2415.

Alers, G.A. and Thompson, D.O. (1961). *J. Appl. Phys.*, 32, 283.

Barnouin, M.J.J. (1972). Observation des configurations des dislocations et mesure de l'énergie de fautes d'empilements, cupro-aluminium. PhD Thesis, CNRS.

Benounich (1990). Mécanique de l'endommagement. PhD Thesis, École d'été d'Oléron/ Presses du CNRS IRSID.

Bohn, R., Klassen, T., Bormann, R. (2001). *Acta Materialia*, 49, 299–311.

Canto, C. (1998). Comportement dynamique de matériaux par déchargements impulsionnels étagés, application au Fer α. PhD Thesis, Nantes University/ISITEM.

Castagné, J.L. (1966). CRSM. *Journal de physique*, Colloque C3, Supplements to nos 7–8.

Chéron, R. and Leroy, M. (1981). *Revue Phys. Appl.*, 36.

Chéron, R., Renaud, J.Y., Leroy, M. (1981). *French Journal of Mechanics*, 77.

Chiem, C.C. (1976). *C. Acad. Sci.*, 282, C-323.

Davies, E.D.H. and Hunter, S.C. (1963). The dynamic compression testing of solids by the method of the split Hopkinson pressure bar. *J. Mech. Phys. Solids*, 11, 155–179.

Dormeval, R. and Stelly, M. (1980). Caractérisation mécanique des matériaux aux grandes vitesses de déformation. Report, CEA R-5044.

Escaig, B. (1968). L'activation thermique des déviations sous faibles contraintes dans les structures h.c. et c.c. Thesis, Paris.

Ferguson, W.G., Kumar, A., Dorn, J.E. (1967). *J. Appl. Phys.*, 38(4), 1863.

Ferguson, W.G., Kumar, A., Dorn, J.E. (1968). *Acta Met.*, 16(1189).

François, D., Pineau, A., Zaoui, A. (1991/1992). *Élasticité et plasticité*. Hermès, Paris.

Friedel, J. (1964). *Dislocations*. Elsevier, Amsterdam.

Gathouffi (1984). Modèle d'endommagement. PhD Thesis, École d'été d'Oléron/Presses du CNRS IRSID.

Guéguen, Y. (1976). Les Dislocations dans l'olivine des péridotites. PhD Thesis, Nantes University.

Guyot, P. and Dorn, J.E. (1967). *Canad. J. Phys.*, 45(983).

Hauser, F.E., Simmons, J.A., Dorn, J.E. (1961). *Response of Metals to High Velocity Deformation*. InterSciences, New York.

Hikata, A. and Elbaum, C. (1967). *Phys. Rev. Letters*, 18, 750.

Horta, R.M., Roberts, W.T., Wilson, D.V. (1970). *Int. J. Mech. Sci.*, 12, 231.

Jaoul, B. (1965). *Étude de la plasticité et application aux métaux*. Dunod, Paris.

Kovács, I. and Zsoldos, L. (1973). *Dislocations and Plastic Deformation*. Pergamon Press, Oxford.

Kumar, A. and Kumble, R.G. (1969). *J. Appl. Phys.*, 40(3475).

Leroy, M. (1970a). Viscoplasticité des métaux cubiques à faces centrées. *C. R. Acad. Sc. Paris, Series C*, 270, 899–902.

Leroy, M. (1970b). Viscoplasticité des métaux cubiques à faces centrées. PhD Thesis, ENSM, Nantes.

Leroy, M. (1972). Sur la viscoplasticité des métaux. Application du formage magnétique. PhD Thesis, Nantes University.

Leroy, M., Cheron, R., Chiem, C.C. (1979). ENSM report. Convention DRET.

Leroy, M., Raad, M.K., Nkule, L., Chéron, R. (1984). Mechanical properties at high rates of strain. In *Institute of Physics Conference*, Bristol, Oxford, 70, 31–38.

Leroy, M., Louvigné, P.F., Longère, P., Nicolazo, C. (1997a). *J. Phys. IV, France*, 7.

Leroy, M., Nicolazo, C., Louvigné, P.F., Thomas, T. (1997b). Dynamic behavior of metals to magnetic field pulses. *EURODYMAT 5th International Conference on Mechanical and Physical Behaviour of Materials under Dynamic Loading*, Oxford.

Lindholm, U.S. (1964). Some experiments with the spilt Hopkinson pressure bar. *J. Mech. Phys. Solids*, 11, 155.

Mason, W.P. (1955). *Phys. Rev.*, 97, 557.

Mason, W.P. (1966). *Phys. Rev.*, 143, 229–235.

Mason, W.P. (1968). *Dislocation Dynamics*. McGraw Hill, New York.

Masumura, R.A., Hazzledine, P.M., Pande, C.S. (1998). *Acta Materialia*, 46, 4527–4534.

McLean, D. (1962). *Mechanical Properties of Metals*. John Wiley and Sons, New York.

Meakin, J.D. and Petch, N.J. (1974). *Phil. Mag.*

Morse, R.W. (1955). *Phys. Rev.*, 97, 1716.

Newey, C.W.A. and Davidge, R.W. (1965). *Dislocations in Lithium Fluorine*. Metallurgical Services & Scientific Techniques Ltd, Betchworth, Surrey.

Otte, H.M. and Hren, J.J. (1966). *Experimental Mechanics*, 6, 177.

Parameswaran, V.R. and Weertman, J. (1969). *Scripta Metallurgica*, 3, 477–480.

Pippard, A.B. (1955). *Phil. Mag.*, 46, 1104.

Renaud, J.Y. (1975). Formage magnétique, courbes limites de déformation. Thesis, École nationale supérieure de mécanique and Nantes University.

Renaud, J.Y. and Leroy, M. (1975). Physique cristalline : comportement viscoplastique (courbes limites de déformation) d'alliages d'aluminium sous l'action de champs magnétiques intenses et pulsés. *C.R. Acad. Sc. Paris*, Series B, 281, 89–92.

Saulnier, A. and Trillat, J.J. (1962). *Le Microscope électronique et la Métallurgie moderne*. Nucléus, Paris.

Seeger, A. (1958). *Kristallplastizität*. Handbuch der Physik/Encyclopedia of Physics, Vol. 2. Springer, Berlin.

Appendix A

Alers, G.A. and Thompson, D.O. (1961). *J. Appl. Phys.*, 32, 283.

Blount, E.J. (1959). *Phys. Rev.*, 114, 418.

Borman, J.A., Wood, D.S., Vreeland, T. (1969). *J. Appl. Phys.*, 40(2), 833.

Cottrell, A.H. (1953). *Dislocation and Plastic Flow in Crystals*. Oxford University Press, New York.

Ferguson, W.G., Kumar, A., Dorn, J.E. (1967). *J. Appl. Phys.*, 38(4), 1863.

Friedel, J.F. (1956). *Dislocations*. Gauthier-Villars, France.

Herve, N. (1971). Étude théorique de l'amortissement des dislocations lors des déformations aux grandes vitesses. PhD Thesis, Nantes University.

Hikata, A. and Elbaum, C. (1967). *Phys. Rev. Letters*, 18, 750.

Hill, B. (1967). *Journal of Mechanics and Physics of Solids*, 15, 2.

Hutchinson, T.S. and Rogers, D.H. (1962). *J. Appl. Phys.*, 33, 792.

Johnston, W.F. and Gilman, J.J. (1958). *Physical Acoustics and Properties of Solids*. D. Von Nostrand Company, Princeton, NJ.

Johnston, W.F. and Gilam, J.J. (1959). *J. Appl. Phys.*, 30, 129.

Kittel, C. (1956). *Introduction to Solid State Physics*. John Wiley and Sons, New York.

Kroner (1958). *Kontinuumstheorie der Versetzungen und Eingenspannungen.* Springer-Verlag, Berlin/Heidelberg.

Lax, E. (1959). *Phys. Rev.*, 115(6), 1591.

Leibfried, G. (1950). *Z. Physik*, 127, 344.

Leibfried, G. and Schlöman, E. (1954). *Nachr. Akad. Wiss Gott.*, 11a, 71.

Leroy, M. (1970a). Viscoplasticité des métaux cubiques à faces centrées. *C. R. Acad. Sc. Paris*, Series C, 270, 899–902.

Leroy, M. (1970b). Viscoplasticité des métaux cubiques à faces centrées. PhD Thesis, ENSM, Nantes.

Leroy, M. (1972). Sur la viscoplasticité des métaux. Application du formage magnétique. PhD Thesis, Nantes University.

Leroy, M. and Offret, S.C. (1972). *R. Acad. Sc. Paris*, vol. 275, Series C.

Lothe, J. (1960). *Phys. Rev.*, 177, 704.

Mason, W.P. (1955). *Phys. Rev.*, 97, 557.

Mason, W.P. (1964). *J. Appl. Phys.*, 35, 2779.

Mason, W.P. (1966). Internal friction measurements and their uses in determining the interaction of acoustic waves with phonons electrons and dislocations. *Phys. Rev.*, 143, 339.

Mason, W.P. and Rosenberg, A. (1966). *Phys. Rev.*, 151, 434.

Mason, W.P. and Rosenberg, A. (1967). *J. Appl. Phys.*, 38, 4.

Parameswaran, V.R. and Weertman, J. (1969). *Scripta Metallurgica*, 3, 477–480.

Pippard, A.B. (1960). *Phil. Mag.*, 46, 1104.

Pope, D.P. and Vreeland, T. (1969). *Phil. Mag.*, 20, 1153.

Stern, R.M. and Granato, A.V. (1962). *Acta Met.*, 10, 958.

Suzuki, T., Ikushima, A., Aoki, M. (1964). *Acta Met.*, 12, 1231.

Taylor, G.I. (1934). The mechanism of plastic deformation of crystals, part 1. In *The Scientific Papers of Sir G.I. Taylor*. Cambridge Royal Society.

Taylor, G.I. (1954). *J. Inst. Civil Engro.*, 26, 486.

White, W. (1954). *Phil. Mag.*, 45, 1343.

Zarka, J. (1967). *Comportement des milieux denses sous hautes pressions dynamiques.* Dunod, Paris.

Zarka, J. (1968). Sur la viscoplasticité des métaux. PhD Thesis, Faculté des sciences de Paris, Paris.

Index

Other titles from

in

Mechanical Engineering and Solid Mechanics

2023

EL HAMI Abdelkhalak, DELAUX David, GRZESKOWIAK Henri
Applied Reliability for Industry 1: Predictive Reliability for the Automobile,
Aeronautics, Defense, Medical, Marine and Space Industries
(Reliability of Multiphysical Systems Set – Volume 16)
Applied Reliability for Industry 2: Experimental Reliability for the
Automobile, Aeronautics, Defense, Medical, Marine and Space Industries
(Reliability of Multiphysical Systems Set – Volume 17)
Applied Reliability for Industry 3: Operational Reliability for the
Automobile, Aeronautics, Defense, Medical, Marine and Space Industries
(Reliability of Multiphysical Systems Set – Volume 18)

2022

BAYLE Franck
Product Maturity 1: Theoretical Principles and Industrial Applications
(Reliability of Multiphysical Systems Set – Volume 12)
Product Maturity 2: Principles and Illustrations
(Reliability of Multiphysical Systems Set – Volume 13)

LEDOUX Michel, EL HAMI Abdelkhalak
Heat Transfer 1: Conduction
(Mathematical and Mechanical Engineering Set – Volume 9)
Heat Transfer 2: Radiative Transfer
(Mathematical and Mechanical Engineering Set – Volume 10)

2020

SALENÇON Jean
Elastoplastic Modeling

2019

BAYLE Franck
Reliability of Maintained Systems Subjected to Wear Failure Mechanisms:
Theory and Applications
(Reliability of Multiphysical Systems Set – Volume 8)

BEN KAHLA Rabeb, BARKAOUI Abdelwahed, MERZOUKI Tarek
Finite Element Method and Medical Imaging Techniques in Bone
Biomechanics
(Mathematical and Mechanical Engineering Set – Volume 8)

IONESCU Ioan R., QUEYREAU Sylvain, PICU Catalin R., SALMAN Oguz Umut
Mechanics and Physics of Solids at Micro- and Nano-Scales

LE VAN Anh, BOUZIDI Rabah
Lagrangian Mechanics: An Advanced Analytical Approach

MICHELITSCH Thomas, PÉREZ RIASCOS Alejandro, COLLET Bernard,
NOWAKOWSKI Andrzej, NICOLLEAU Franck
Fractional Dynamics on Networks and Lattices

SALENÇON Jean
Viscoelastic Modeling for Structural Analysis

VÉNIZÉLOS Georges, EL HAMI Abdelkhalak
Movement Equations 5: Dynamics of a Set of Solids
(Non-deformable Solid Mechanics Set – Volume 5)

2018

BOREL Michel, VÉNIZÉLOS Georges
Movement Equations 4: Equilibriums and Small Movements
(Non-deformable Solid Mechanics Set – Volume 4)

FROSSARD Etienne
Granular Geomaterials Dissipative Mechanics: Theory and Applications in
Civil Engineering

RADI Bouchaib, EL HAMI Abdelkhalak
Advanced Numerical Methods with Matlab® 1: Function Approximation
and System Resolution
(Mathematical and Mechanical Engineering Set – Volume 6)
Advanced Numerical Methods with Matlab® 2: Resolution of Nonlinear,
Differential and Partial Differential Equations
(Mathematical and Mechanical Engineering Set – Volume 7)

SALENÇON Jean
Virtual Work Approach to Mechanical Modeling

2017

BOREL Michel, VÉNIZÉLOS Georges
Movement Equations 2: Mathematical and Methodological Supplements
(Non-deformable Solid Mechanics Set – Volume 2)
Movement Equations 3: Dynamics and Fundamental Principle
(Non-deformable Solid Mechanics Set – Volume 3)

BOUVET Christophe
Mechanics of Aeronautical Solids, Materials and Structures
Mechanics of Aeronautical Composite Materials

BRANCHERIE Delphine, FEISSEL Pierre, BOUVIER Salima,
IBRAHIMBEGOVIC Adnan
From Microstructure Investigations to Multiscale Modeling:
Bridging the Gap

CHEBEL-MORELLO Brigitte, NICOD Jean-Marc, VARNIER Christophe
From Prognostics and Health Systems Management to Predictive
Maintenance 2: Knowledge, Traceability and Decision
(Reliability of Multiphysical Systems Set – Volume 7)

EL HAMI Abdelkhalak, RADI Bouchaib
Dynamics of Large Structures and Inverse Problems
(Mathematical and Mechanical Engineering Set – Volume 5)
Fluid-Structure Interactions and Uncertainties: Ansys and Fluent Tools
(Reliability of Multiphysical Systems Set – Volume 6)

KHARMANDA Ghias, EL HAMI Abdelkhalak
Biomechanics: Optimization, Uncertainties and Reliability
(Reliability of Multiphysical Systems Set – Volume 5)

LEDOUX Michel, EL HAMI Abdelkhalak
Compressible Flow Propulsion and Digital Approaches in Fluid Mechanics
(Mathematical and Mechanical Engineering Set – Volume 4)
Fluid Mechanics: Analytical Methods
(Mathematical and Mechanical Engineering Set – Volume 3)

MORI Yvon
Mechanical Vibrations: Applications to Equipment

2016

BOREL Michel, VÉNIZÉLOS Georges
Movement Equations 1: Location, Kinematics and Kinetics
(Non-deformable Solid Mechanics Set – Volume 1)

BOYARD Nicolas
Heat Transfer in Polymer Composite Materials

CARDON Alain, ITMI Mhamed
New Autonomous Systems
(Reliability of Multiphysical Systems Set – Volume 1)

2014

ATANACKOVIC M. Teodor, PILIPOVIC Stevan, STANKOVIC Bogoljub, ZORICA Dusan
Fractional Calculus with Applications in Mechanics: Vibrations and Diffusion Processes
Fractional Calculus with Applications in Mechanics: Wave Propagation, Impact and Variational Principles

CIBLAC Thierry, MOREL Jean-Claude
Sustainable Masonry: Stability and Behavior of Structures

ILANKO Sinniah, MONTERRUBIO Luis E., MOCHIDA Yusuke
The Rayleigh–Ritz Method for Structural Analysis

LALANNE Christian
Mechanical Vibration and Shock Analysis – 5-volume series – 3rd edition
Sinusoidal Vibration – Volume 1
Mechanical Shock – Volume 2
Random Vibration – Volume 3
Fatigue Damage – Volume 4
Specification Development – Volume 5

LEMAIRE Maurice
Uncertainty and Mechanics

2013

ADHIKARI Sondipon
Structural Dynamic Analysis with Generalized Damping Models: Analysis

ADHIKARI Sondipon
Structural Dynamic Analysis with Generalized Damping Models: Identification

BAILLY Patrice
Materials and Structures under Shock and Impact

BASTIEN Jérôme, BERNARDIN Frédéric, LAMARQUE Claude-Henri
*Non-smooth Deterministic or Stochastic Discrete Dynamical Systems:
Applications to Models with Friction or Impact*

EL HAMI Abdelkhalak, RADI Bouchaib
Uncertainty and Optimization in Structural Mechanics

KIRILLOV Oleg N., PELINOVSKY Dmitry E.
Nonlinear Physical Systems: Spectral Analysis, Stability and Bifurcations

LUONGO Angelo, ZULLI Daniele
Mathematical Models of Beams and Cables

SALENÇON Jean
Yield Design

2012

DAVIM J. Paulo
Mechanical Engineering Education

DUPEUX Michel, BRACCINI Muriel
Mechanics of Solid Interfaces

ELISHAKOFF Isaac *et al.*
*Carbon Nanotubes and Nanosensors: Vibration, Buckling
and Ballistic Impact*

GRÉDIAC Michel, HILD François
Full-Field Measurements and Identification in Solid Mechanics

GROUS Ammar
*Fracture Mechanics – 3-volume series
Analysis of Reliability and Quality Control – Volume 1
Applied Reliability – Volume 2
Applied Quality Control – Volume 3*

RECHO Naman
Fracture Mechanics and Crack Growth

2011

KRYSINSKI Tomasz, MALBURET François
Mechanical Instability

SOUSTELLE Michel
An Introduction to Chemical Kinetics

2010

BREITKOPF Piotr, FILOMENO COELHO Rajan
Multidisciplinary Design Optimization in Computational Mechanics

DAVIM J. Paulo
Biotribology

PAULTRE Patrick
Dynamics of Structures

SOUSTELLE Michel
Handbook of Heterogenous Kinetics

2009

BERLIOZ Alain, TROMPETTE Philippe
Solid Mechanics using the Finite Element Method

LEMAIRE Maurice
Structural Reliability

2007

GIRARD Alain, ROY Nicolas
Structural Dynamics in Industry

GUINEBRETIÈRE René
X-ray Diffraction by Polycrystalline Materials

KRYSINSKI Tomasz, MALBURET François
Mechanical Vibrations

Printed and bound by CPI Group (UK) Ltd, Croydon, CR0 4YY

16/04/2025

14658458-0003